SOCIETY IN TIME AND SPACE

A geographical perspective on change

Society in Time and Space is an important and innovative book which offers a geographical perspective on societal change, and sets out to show how understanding the geography of such change enables us to appreciate better the basic processes involved. Robert Dodgshon argues that, as a first step, we need to clarify the circumstances under which society becomes inertial and finds change difficult. Using a range of historical and contemporary examples, he shows that society's use of space is a powerful source of this inertia. Different sources of geographical inertia are explored, including society's symbolization and organizational structuring of space, together with its capitalization of landscape. Building on this mapping of inertia, Professor Dodgshon shows how society has long steered radical change around such space.

Society in Time and Space will be of interest not only to geographers but also to historians and social theorists.

ROBERT A. DODGSHON is Professor of Human Geography at the University of Wales, Aberystwyth. His other books include, as author, *The European Past* (1987) and *From Chiefs to Landlords* (1998) and, as co-editor, *An Historical Geography of England and Wales* (1978 and 1990) and *An Historical Geography of Europe* (1998).

Cambridge Studies in Historical Geography 27

Cambridge Studies in Historical Geography encourages exploration of the philosophies, methodologies and techniques of historical geography and publishes the results of new research within all branches of the subject. It endeavours to secure the marriage of traditional scholarship with innovative approaches to problems and to sources, aiming in this way to provide a focus for the discipline and to contribute towards its development. The series is an international forum for publication in historical geography which also promotes contact with workers in cognate disciplines.

For a full list of titles in the series, please see end of book.

SOCIETY IN TIME AND SPACE

A Geographical Perspective on Change

ROBERT A. DODGSHON

CAMBRIDGE
UNIVERSITY PRESS

PUBLISHED BY THE PRESS SYNDICATE OF THE UNIVERSITY OF CAMBRIDGE
The Pitt Building, Trumpington Street, Cambridge CB2 1RP, United Kingdom

CAMBRIDGE UNIVERSITY PRESS
The Edinburgh Building, Cambridge CB2 2RU, United Kingdom
    http://www.cup.cam.ac.uk
40 West 20th Street, New York, NY 10011–4211, USA
    http://www.cup.org
10 Stamford Road, Oakleigh, Melbourne 3166, Australia

First published 1998

Printed in the United Kingdom at the University Press, Cambridge

Typeset in Times NR 10/12 pt   [SE]

A catalogue record for this book is available from the British Library

ISBN 0 521 59385 9 hardback
ISBN 0 521 59640 8 paperback

Contents

List of figures   *page* ix
List of tables   x
Preface   xi

1   Human geography and societal change   1
2   Establishing a taxonomy of societal change   21
3   The experience of change: world systems and empires   51
4   The experience of change: states and regions   84
5   Sources of inertia: the cultural construction of landscape   104
6   Organizational sources of inertia   123
7   The built environment as a source of inertia   139
8   Conceptualizing inertia: the first step towards a geography of
    societal change   162
9   The geography of societal change: a concluding synthesis   181

References   202
Index   223

Figures

3.1 How core and peripheral areas differ within a capitalist world
   system   *page* 55
3.2 Mesopotamian civilizations and empires   57
3.3 Changing configurations of the Chinese Empire   62
3.4 The dynamics of a capitalist world system   73
7.1 Fixed reproducible assets and national wealth, Great Britain
   1760–1860 (at constant prices, 1851–60)   152
7.2 Fixed reproducible assets and national wealth, United Kingdom
   1850–1920 (at constant prices, 1900)   152
7.3 Net stock of fixed reproducible assets by sector 1760–1860 in
   Great Britain (at constant prices, 1851–60)   153
7.4 Net stock of fixed reproducible assets by sector 1850–1920 in
   the United Kingdom (at constant prices, 1900)   153

Tables

2.1 The dynamics of societal systems   *page* 23
2.2 Sources of societal change   31
2.3 The product of societal change   45
2.4 The morphology of societal change   48

Preface

In origin, this book arises from my longstanding interest in the institutional basis of landscape development. Its precise roots lie in work which I first carried out over twenty years ago on how far-reaching changes in medieval land tenure exploited space strategically by working their way around established customary patterns of tenure. Subsequent work made me conscious of how the broad principle involved, the opportunistic use of space to develop new institutional forms, could be developed at other scales and in a wide variety of other contexts. Examples of such change were discussed in my *European Past* (1987) book, including a detailed case-study of how the decline of feudalism could be analysed in such terms. The aim of this book is to develop this geographical perspective on change in a more direct and systematic way. It will address three broad themes. First, I want to provide a more extended synthesis of historical examples and to show how they feature in debates that range across social, political, economic and cultural change. Second, I want to explore how the foundations for the strategic use of space by society derive from society's tendency to become inertial *through its use of space*. Third, I want to explore the problems posed by trying to conceptualize this inertia and its effects on the geography of societal change.

Though the book is a study in large-scale synthesis, or 'big-picture' history, from the perspective of historical geography, it is not written against or as a challenge to historical geography's abiding concern for a more reductionist methodology or its supposedly new-found concern for the situatedness of all knowledge. Rather is it my conviction that historical geography should operate at different scales, even if the different scales involve different problems and methodologies. Indeed, in its final stages, this book has been produced alongside another that deals with my research on the western Highlands and Islands of Scotland during the early modern period and which takes a more reductionist approach.

The drafting and redrafting of the text has been drawn out over a number of years, but the first really acceptable version of the manuscript benefited from research leave granted during 1995–6. I am grateful to my former Principal and Vice-Chancellor, Professor Kenneth O. Morgan, for his support with this leave. I am also extremely grateful to two anonymous referees, and to one of the series editors, Dr Alan Baker, for their valuable comments on drafts of the book. Thanks are also due to my departmental cartographers, Ian Gulley and Anthony Smith, for drawing the diagrams. As with my other books, my wife, Katherine, has provided the kind of support that goes beyond in-text citations and references.

Over the past decade, parts of the book's argument have been presented in various guises to seminars at Bristol, Cambridge, Cork, Exeter, Liverpool, London and Lund. I am grateful to all those whose questions and comments served continually to remind me that large open-ended topics like societal change are really intellectual journeys without hope of a final destination. Rather must we be content with developing insights into those landscapes through which the debate may pass more slowly.

1

Human geography and societal change

> . . . if spatial organization makes a difference to how society works and how it changes, then far from being the realm of stasis, space and the spatial are also implicated (*contra* Laclau) in the production of history – and thus, potentially, in politics
>
> (Massey 1992: 70)

This is a book about the geography of societal or socio-cultural change. By the terms societal or socio-cultural, I mean society in its wider sense, that is, as a compound of its cultural, social, political and economic forms. Throughout, I will use the terms societal or socio-cultural to capture this wider generic meaning and the terms cultural, social, political and economic when I want to refer to changes that are rooted in more specific aspects of society's character. However, this clarification of terminology is not without its problems when reviewing the debate over the nature of society and how it changes. Social theorists, within and without geography, have tended to use terms like social change generically, as a cover for all aspects of society's constitutive character. Yet the way in which the debate over what matters most, or what is determinate, has shifted, with current views favouring a cultural turn, is itself an argument for a terminological usage that is able to differentiate between the whole and its parts, even if some parts are seen as capable of shaping or grounding the whole.

In addressing the geography of societal change, I do not mean the descriptive study of where change has taken place or the reconstruction of particular instances of change within their geographical setting. That change happens somewhere and has a geography is self-evident and hardly needs stating. To say that we are interested in societal change because it produces new geographies is also self-evident. Rather is my objective to show how understanding this geography can contribute to a concept of how society changes, one that helps to explain how and why change tends to occur where it does. In seeking to develop such an understanding, I want to build on the assumption that the geography of societal processes is not some minor side-show, of interest only to geographers. As an increasing number of social theorists and geographers have claimed, the 'spatiality' of society, the way in which it uses space, matters fundamentally to the very nature of society and its constitutive processes. 'There are', declares Soja, 'no aspatial social processes' (Soja 1996: 46; Soja 1989). We live in a world that is only meaningful through the fact that it is tangible, built around concrete experiences and situations. But equally, it is only meaningful through the fact that it is spatially as well as temporally extended, an accumulation of spaces as much as it is an accumulation of moments. I want to suggest that seeing the geography of societal change as contributing to a wider concept of change is a logical extension of this debate about the spatiality of social life. As an agenda, it requires us to consider how far the process of societal change is articulated geographically and, following on from this, whether there are types of change that use space strategically so that their analysis in geographical terms provides an essential basis for their understanding.

The retreat from structure

In so far as human geography has addressed the wider problem of societal change, it has shown little real conviction over producing a geographically sensitive reading of the problem, one that binds geography into the processes of change. If we look at the branch of the discipline whose subject matter should be shot through with such insights, historical geography, we find that it has actually engaged the problem in an uneven and selective way. Arguably, the exemplary work of writers like Sauer (1952) and Meinig (1986 and 1993) on 'big-picture' themes has not been matched by an entrenched and maturing interest in such themes amongst historical geographers at large. Meinig's *Shaping of America* especially, has much to say on matters of socio-cultural change, generally as well as specifically. In Britain, Darby's work on the mapping of the Domesday Book also rates as a project of a comparable scale and vision (Darby 1977). However, with its emphasis on the use of large-scale data sets to reconstruct past geographies, Darby's methodology tended to reduce historical geography to a succession of static images with little of real significance for the wider debate over societal change. To be fair, Darby himself argued that these cross-sectional approaches should be linked via comparative analyses that both highlighted and explained the changes in between (Darby 1962: 127–56), but the dynamic introduced by such comparative analyses did easily melt the somewhat frozen detail out of which his geographies were produced.

Yet whilst it could be argued that historical geographers generally have shown only limited interest in large-scale syntheses, they cannot be accused of lacking interest in societal change from a conceptual or empirical point of view. An increasing number of studies have addressed specific aspects, but methodologically, their approach has largely been close-focused, stressing, albeit in different ways, the situatedness of knowledge about change. Some have been largely text-based, engaging in the discourse over themes like modernity (Gregory 1990a: 217–33; Gregory 1991: 17–44; Harris 1991: 671–83) or changes in the nature of power (Philo 1991: 137–61; Driver 1992: 147–56) through the work of particular scholars and stressing how interpretations are socially constructed. Others have been place-based, researching local or regional instances of change through a powerful blend of theory and data. Langton and Höppe's work on early Swedish industrialization well illustrates this sort of approach (Langton and Höppe 1995). Though they are amongst those who have also considered change on a much larger scale (Langton and Höppe 1983), their stress on the need to see early Swedish industrialization through the recovery and specification of local time–space geographies exemplifies the close-focus approach of much that has been written in historical geography over recent years. In fact, despite the far-reaching shifts in the subject matter of historical geography over the 1980s and 90s, and a growing concern for concepts, much of the drift of recent work has been to reinforce rather than weaken the primacy of reductionist methodologies and to stress the problems attached to large-scale, overarching generalizations (see, for example, Butlin 1993; Philo 1994: 252–81; Baker 1996: 1–24; but see Baker 1982: 233–43). Of course, in this respect, we are simply acknowledging that many historical geographers have been influenced, knowingly or unknowingly, by post-modernist thinking and the turn against metatheory. In this sense, Gregory's recent call for a greater awareness of how a geography of the body and personal space can be linked in with the wider notion of spatiality, for recognizing 'the corporeality of vision', appears a natural extension of how historical–geographical discourse has moved in recent years (Gregory 1994: 416).

The shift towards a more reductionist and personalized perspective is of course part of a wider trend within human geography. The high-water mark of human geography as a would-be/could-be science based on some form of spatial analysis has long passed. Indeed, relatively few would now defend a view of human geography that seemingly dehumanized landscape by reducing it to a patterning of equilibrated systems and gravity formulations that had human interaction decaying in a perfectly graduated and normative way as one moved away from centres of population. Nothing better illustrates this shift than the extent to which those who pioneered such an approach with so much distinction in the 1950s and 60s have themselves been instrumental in trying to re-people human geography's landscapes of study. The economic geographer, Berry, is amongst those who can now be met on the road from Damascus. In a recent overview of economic geography, he has called for it to be reformulated so as to incorporate a cultural component, a greater sensitivity to how different cultures construct their own reading of economic geography.

Explicating this cultural component, he suggests, must be 'the first step towards the reconstruction of economic geography's conceptual core' (Berry 1989: 18). In step with what has happened in human geography as a whole, Berry has shifted his analysis from the study of physical structures, networks and systems in their own terms, to their study as socio-cultural constructions.

Geographers have a double problem here, for the assertion that 'every object' is 'an object for a subject' (Ley 1977: 498) applies as much to their own vision of the world as to the visions of those whom they study. Instead of a human geography built around a simple or primary observer–object relationship, or around how the geographer observes and records the world, it has become a discipline more concerned with secondary observer–object relationships, dealing with how the geographer observes other people observing, categorizing or experiencing the world, a shift in emphasis which compares with the emic/etic debate in anthropology (Harris 1979: 32–4). Once we admit this degree of relativity, with all its problems of 'double translation' (Olsson 1982: 223) or – to use a geographical metaphor for the potential miscegenation of meaning –'travelling theory' (Gregory 1994: 9–14), the notion of geography as an objective or stable science fades. In fact, for some, 'unmasking' geography's 'pretensions of objectivism' (ibid.: 86) is not an incidental consequence of recent shifts in perspective, but its primary goal.

Inevitably, this permeation of human geography by relativism has been accompanied by a revaluation of the role played by structure or determination in geography. Gregory has played an instrumental role in driving this debate and has provided the most comprehensive and sustained critique of geographical structure. In part, he bases his rejection on moral grounds. The understanding of societal systems and structures in terms of imperatives and constraints provides the basis for a system of social control. However, it is also a question of concept. He rejects such approaches because of their inability to handle the purposiveness of social systems and 'the semantics and sociology of discourse' (Gregory 1980: 335). For Gregory, a critical human geography deals with 'knowledgeable' or reflexive beings rather than structures and systems that function in silence. Influenced by Habermas's proposition that societal change is, in the first instance, about changes in social consciousness, that is, about changes in intersubjectivity and communicative rationality, Gregory believes that no amount of tinkering with systems theory can embody the nuances of meaning and richness of values embodied in such dimensions of social consciousness.

A close reading of his work suggests a progressive and deepening emasculation of structure as an independent source of determination for society. In his early writing, he clearly found Thompson's notion of a 'bounded human agency' helpful to his thinking, conceding that 'a concept of determination has to be allowed back in; but in terms which return us once again to the capacities and capabilities of a bounded human agency' (Gregory 1981: 13).

Amongst available methodologies, he has some sympathy with structuration theory because it treats the properties of social systems as both the medium and the outcome of the practices that constitute them. By linking structure and practice together in this way, structure ceases to be beyond the reach of everyday social practice, with its own logic and history. In the process, it also ceases to be 'a constraint or a barrier to action but is instead essentially involved in its reproduction' (ibid.: 10). Latterly, though, Gregory has defined his position more starkly. Whereas Giddens juxtaposed structure 'as recursively organized sets of rules and resources' existing 'out of time and space' with system as 'the situated activities of human agents, reproduced across time and space' (Giddens 1984: 25), Gregory finds the former too typological and abstract, preferring instead to give more emphasis to the latter: the individual settings and instances of social reproduction (Gregory 1994: 123). He aligns himself here with Mann. His restatement of Mann's assertion that the history of power is the 'history of particular places' (Mann 1986: 40) could just as easily be paraphrased as the history of all geography is the history of particular places or of the 'complexities of particular places' as Harris put it (Harris 1991: 681). Seemingly, for Gregory, there are no hidden structures, or rules governing such structures, waiting to be discovered. Herein lies a vital disagreement with Giddens. For both, all social life is contingent and contextual, but whereas Giddens accepts that there may be hidden organizing rules to be discovered that apply out of time and space, Gregory feels that the study of each particular place or locale is sufficient unto itself and has no need of wider reference.

In seeking to suppress any autonomous or paramount role for structure, Gregory's aim is the wholly laudable one of releasing human geography's humanist potential, unfettered by structural and material constraints or ready-packaged as a study in control systems for those in power. Towards this end, he has long argued in principle and by example that human geography should be grounded in social theory. As well as making the role of structure more transparent, this has had the effect of narrowing – one might even say, constraining – the conceptual material out of which its discourse is now woven. From being a subject that is richly grounded in all areas of the social sciences, it threatens today to become rooted solely in social theory *sensu stricto*. Indeed, some would claim that instead of standing in geography, and thinking about social theory, we should reposition ourselves so as to stand in social theory and think geographically.

Social theory and geography

It is crucial to any understanding of how social theory has colonized human geography to realize that the discovery, like all true discoveries of one culture by another, was mutual. Social theorists began incorporating geography into their thinking, a flattering of the spatial perspective which has encouraged geographers to reciprocate. Two social theorists have been especially influential: Giddens and Lefebvre.

As already indicated, Giddens tames the freedom or autonomy of structure by using structuration theory, but it can hardly be said that he dispenses with structure. He acknowledges the traditional definition of structure as something that has formative or architectural qualities, and which, being 'external' to human action, acts 'as a source of constraint' on the individual (Giddens 1984: 16). By comparison, his own treatment of structure seeks to bring it within reach of social practice by treating it as part of a duality whereby structure and everyday social practice are linked recursively, each being both the medium and outcome for the other. Seen in this way, structure becomes 'reproduced social practices' that exist as 'time–space presence, only in its instantiations in such practices and as memory traces orienting the conduct of knowledgeable human agents' (ibid.: 17). Around these core ideas, however, Giddens uses structure freely and in ways that give it a potent geographical meaning. The key to this geographical meaning lies in his concept of time–space distanciation, or 'the stretching of social systems across time–space' incorporating and integrating presence and absence (ibid.: 181). Seen in relation to the notion of time–space distanciation, structure is not something tangible in itself but comprises 'the structuring properties' that help bind social systems in time–space via social practice and which enabled 'similar social practices to exist across varying spans of time and space and which lend them "systemic" form' (ibid.: 17). In short, structure is made up of 'principles of organization' that sustain 'recognizably consistent forms of time–space distanciation' (ibid.: 181). The main structural principles derive from what he calls the structures of domination around which society and its social practices are ordered: these can be broken down into the 'containers' that 'store allocative and authoritative resources' (ibid.: 262). Again, implicit in these principles, is the fact that such resources underpin and sustain the time–space distanciation of a society.

Many human geographers have responded positively to Giddens's theory of structuration. However, I want to confine myself to Pred's handling of the concept for he uses it so as to produce a geographically framed interpretation of change that is particularly insightful, one that must be confronted in any reconceptualization of the problem. Its power as a reworking stems from the way he tries to align structuration theory with the concepts of time–geography. With its stress on duality rather than dualism, structuration theory enables the analysis of individual life-paths, with their mix of possibility and constraint, to be bound in a reciprocal way to the institutional projects around which time–geography is organized: the former representing everyday social practices and the latter, structure. In his own words, if 'all society's formal and informal institutions are project-bound' then the 'detailed situation and material continuity of structuration are perpetually spelled out by the inter-section of individual paths with institutional projects occurring at specific temporal and spatial locations' (Pred 1984: 281–2). Pred turns this statement about the ongoingness of social practices and structures into a concept of change through the way in which he develops the notion of instantiation, or the fact that social structures exist only in their implementation via social practices in specific time–space situations (ibid.: 280–1). Because of the way different social practices intersect or are coupled through each instantiation, they can be said to produce place or geography. At the same time, because the social practices that exist at each moment are a product of past experience, 'a previously sedimented array of cultural and social practices' (ibid.: 285), and because of the way in which they all interact via life-paths and institutional projects, they have the potential both to reproduce and transform each other, both as practices and as structures (ibid.: 281; Pred 1985: 343). For this reason, Pred presents place and change as inextricably linked. Places exist only through the constant reproduction of their social and cultural forms via situ-ated social practices. Through this reproduction, Pred sees each place as con-stituted through a process of becoming, always having the potential to change. Yet though place is constituted through a process of becoming, this process of becoming is historically contingent, constrained by past experience and the situated or prior social practices through which it works itself out.

Like Giddens, Pred is inclined to convert structure into a digestible form, to refine it, in order that it can be reproduced through social practices and life-paths. We can see this in his treatment of structure. In keeping with much that is being written in human geography at present, he sees power relations as being at the heart of social structure (Pred 1985: 339). Following Foucault, these are seen by Pred as inseparable 'from the realm of action and everyday practices' (ibid.: 339). The subjugation of power relations to everyday practice though, is more easily assumed than demonstrated. Whilst being given sufficient power of determination to succeed at being hegemonic, they are also sufficiently embedded or ensnared within day-to-day practice to be repro-duced and transformed by it. How much needs to be accommodated is well shown by Pred's handling of language. At one and the same time, language is one of the most innovative yet most inertial of cultural traits. But faced with an interpretation that draws everything on board, Pred sees the transforma-tion of language and its associated codes as bound up with institutional pro-jects and as part of the 'process whereby power relations become practices and practices become power relations' (ibid.: 340). Clearly, there is little scope for cultural inertia or lags here, language being entirely internalized within the social process of the moment in an ongoing process of becoming.

Another influential figure in human geography's recent search for a socio-theoretic base has been the French social theorist, Lefebvre. Like Giddens, he has devoted much thought to how space should be incorporated into a theory of society. The central tenet of his thinking on the matter is that space is not given or outside of social practices and their performance, but is produced through social practice itself. All space, in other words, is social space, a social product made out of a raw material that is Nature (Lefebvre 1991: 84). At first, the character of this produced space was 'biomorphic and anthropological' (ibid.: 229), but in time, and especially with the rise of capitalism, its character became less immediate. It acquired 'a sort of reality of its own', associated with but distinct from the commodities and capital of the global economy (ibid.: 26). It also became 'a means of control, and hence of domination; of power' (ibid.: 26), a landscape configured by the social relations of capitalism.

Stated thus far, his ideas offer little that would be considered particularly insightful to human geographers given what has been written over the past decade, with or without Lefebvre. However, his interpretation has other more critical dimensions to it. First, and of direct interest to the central theme of this book, is his suggestion that whilst landscapes and places are seen as having a vital inheritance of forms and practices, 'space is always, now and formerly a *present* space, given an immediate whole, complete with its association and connections in their actuality' (ibid.: 37). This supposed erasure of space's compounded past is a theme in Lefebvre's work that I will respond to later. Second, and crucial to the inner logic of his argument, 'production process and product present themselves as two inseparable aspects, not as two separable ideas' (ibid.: 37). This inseparability has this consequence for how we see space. Because it is the outcome of a process, a sequence of operations, it cannot be treated as 'a simple object' (ibid.: 73). This is a critical insight, one that has been commented on by Merrifield. The production of Lefebvrean space, he argued, comprises the process as well as the outcome. Referring to the sequences and flows that produced it, and the way in which flows of capital and resources undergo 'thingification' around particular points in the form of fixed capital, Lefebvrean space becomes 'the totality of the flow and thing' (Merrifield 1993: 521 and 525). Of course, this has the effect of binding space into social theory as part of its problematic. A third influential feature of Lefebvre's conceptualization of space is the way in which he explored capitalism's capacity to differentiate space and to create tensions and conflicts within it, notably that between the global and what he calls the sub-divided or fragmented, a contradiction that embraces what would conventionally be seen as core–periphery conflicts. It was through these tensions and conflicts that he felt space was assuming a greater role in society, a view strongly buttressed by his belief that capitalism's production of space in various forms was in part responsible for 'the survival of capitalism' (Lefebvre 1991: 346). A fourth theme in Lefebvre's thought is really a subset of the previous theme but deserves to stand on its own. It is Lefebvre's suggestion, one taken up and much developed by Harvey and others, that the overall nature of capitalism has changed, with the arena of urbanization now being more important to capitalism than industrialization *per se* (ibid.: 321–40; Harvey 1985: 185–226; Soja 1989: 97).

Lefebvrean concepts of space have been widely utilized by geographers. Harvey's work has been especially significant in diffusing them amongst geographers. Yet whilst acknowledging that we owe the broad concept of space 'as social power' to Lefebvre (Harvey 1989: 226), Harvey has not hesitated to rework them through his own reading of space within a capitalist framework. Arguably, he has taken Lefebvre's rather coarse shift of capital from a phase of industrialization to a phase in which urbanization is more important and has produced a more fine-textured analysis of how capitalism has been phased in its strategic use of space, drawing out more clearly not only the cyclic switches that can take place between the primary and secondary flows of capital, but its recent phasing in terms of Fordism, post-Fordism and flexible accumulation. In doing so, he has given a different emphasis to two aspects of the argument. First, it is difficult when reading Harvey's analysis of capital, and the tension created by the conflict between mobile and fixed capital, not to be left with the impression that he has not wholly accepted Lefebvre's notion of space as both process and product. The very essence of the contradiction involved is that capital needs to be mobile, yet cannot recover that part of itself already sunk into the fixed capital. When Harvey talks about this being not just a source of tensions but of 'contradictions' (Harvey 1978: 124; Harvey 1985: 24–5), he is surely arguing that however much capital, and its embedded social relations, may wish to rework landscape as process, what is of the essence to capitalism is that it does not wholly succeed. In other words, there is a part of space that remains ineluctably a product, a spent decision over how resources of space should be used. This leads directly to my second point which is that compared to Lefebvre, Harvey seems to favour a different position on how time and space relate to each other. Certainly, Lefebvre does not neglect time, nor can he be said to ignore how it substitutes for space. At one point, he asked what does a buyer get when he purchases a space, and answered it by saying time (Lefebvre 1991: 356). As the passage quoted earlier shows, though, he seemed inclined to collapse time into the flatter, singular form of synchronic time, with any inherited processes being treated as part of a single totality irrespective of the rate of turnover or change at which they are moving. Harvey meanwhile, through his declared project to reconstruct the historical geography of capitalism, offers a more developed view of how capitalism moves through time and of how time–space compression has accentuated its tensions (i.e. Harvey 1989: 211–59), trends which he sees as so fundamental to the modern world that he calls for the 'positioning [of] our geography between space and time' (Harvey 1990: 433). Arguably, such a view can only be meaningful if it allows for the experience of time to be variable, or produced socially just like space (Harvey 1989: 223–5), so that each moment in space is potentially diachronic in terms of the different relations and processes that it brings together as a simultaneous experience (cf. Massey 1992: 81).

Lefebvre's particular treatment of time does, in fact, raise a further, more general point, one that is intimately linked to the growing emasculation of structure as a factor in socio-cultural change. As part of his attempt to raise the significance of space within social theory, he tended to suppress the role of historical time or historicity. 'Points and systems of reference inherited from the past', he proclaimed, 'are in dissolution' (Lefebvre 1991: 416). Their 'values crumble and clash' (ibid.: 417). In a phrase that strikes at the very core of his argument, he argued that all past values and historical forms are now subjected to 'a *trial by space*', one that sifts and sorts them spatially (ibid.: 417–18). Taken on its own, this last point is hardly contentious. However, his suggestion that the past is *currently* undergoing 'a *trial by space*' contains a much more radical assertion. Put simply, he sees the link between the past and present, at least in today's terms, as being attenuated or, to use his own phrase, 'in dissolution': so much so that it amounts to something of a disjuncture. Others have echoed the rather surgical way in which he would now have us separate past from present. Jameson, for instance, claims the present as now 'autoreferential' (Jameson 1991: xii) and that there has been a break in the 'signifying chain' between past, present and future (ibid.: 127), with time reduced to 'a series of pure and unrelated presents in time' (ibid.: 27). The very essence of post-modernism, he argues, is 'a situation in which the survival, the residue, the holdover, the archaic, has finally been swept away without a trace' (ibid.: 309), leaving the present almost as a *non sequitur*. Bhabha too, speaks of a 'disjunctive present' (Bhabha 1994: 254), but his purpose is slightly different, being to challenge the notion that the history or narrative of culture can be written in 'homogeneous, serial time' (ibid.: 37). Rather is culture being constantly driven across the differences that come from what he labels as the 'enunciative split' between *énoncé* and enunciation, a split that admits cultural differences through a continual translation of meaning and which thereby destroys 'the logics of synchronicity and evolution' (ibid.: 36–7 and 164). This supposed instantaneity of experience within post-modern or, for Bhabha, post-colonial society, is bound up with the weakening of structure as a determining factor in socio-theoretic discourse, for the claimed disjuncture between past and present signals the capacity of social process to remake the world *and its structures* afresh, to provide it with an enunciation that is continually of the moment.

Geography and the retrieval of structure

In turning to a more direct critique of these recent trends in human geography and associated readings of spatiality, I want to make it clear that I see the debate on whether human geography should be grounded in behavioural, cognitive or social-theoretic approaches as having yielded fundamental gains for its methodological and philosophical development. But as a counter-argument, I would suggest that their considered exclusion or suppression of structure as something to be studied in its own terms has led to the neglect of other equally valid and fruitful viewpoints (cf. Lloyd 1993: 6–7). Indeed, I want to argue that there can be no realistic or critical concept of societal change, whether developed from within a geographical perspective or outside it, without accepting a more constraining role for structure.

To this end, the human geographer needs to take a more defensive stance over structure. The current problem centres on the widespread use now made of structuration theory or its derivatives by geographers. As a dimension of societal change, structure or organization does not have an agreed definition. Potentially, it can be defined in a variety of different ways but once we adopt structuration theory as the basis of our methodology, we are effectively adopting a particular, even narrow, interpretation of structure. This is because if structure is to be brought within reach of everyday social practice, its character has to be refined or reduced to what is reproducible as social practice. This comes across from Giddens's treatment of it. In different contexts, he accepts the importance of both deep-rooted cognitive structures which might lie beyond any self-interrogation and the material forms through which social systems help organize their allocative and authoritative resources. Yet when it comes to his operating definition of structure, he is forced to dematerialize it, stressing soft- at the expense of hard-programmed structures. Structural roughage, including the bricks and mortar or fixed capital that give so much capacity and extension to modern social, economic and political structures, is excluded. Intrinsically geographical but materialist concepts of structure, like adaptation, are dismissed out of hand as 'vacuous' concepts (Giddens 1984: 233–5), with structure being variously reduced to 'structural principles', 'rules' and 'memory traces'. In effect, in order to define a structure that can be carried by or reproduced through everyday social practice, it becomes a definition of structure merely as hand baggage.

Though such a stripped-down idea of structure is now widely articulated in human geography, a sceptic might argue that it is guided not so much by a limited metaphor as by a false one. Only select aspects of structure are actually 'reproduced' through individual instantiations of structural principles and their everyday social practice. Some might go further and argue that as a mechanism of change, structuration confuses the process of reproduction with that of evolution. By its very character, many key aspects of structure are sustained or transformed in ways that structuration theory cannot handle or for which it is an inappropriate form of conceptualization, a point not lost on critics of structuration theory (Gregson 1986: 197–8). Indeed, I want to suggest that the essence of structure, and its potential role as a factor within any concept of social or geographical change, lies in the fact that everyday social practice does not embrace structure in all its diverse forms, at least not so easily as notions of structuration would suppose.

By calling for a wider, more open definition of structure here, I am not simply proposing that there are aspects of social structure that are not easily or comfortably contained within the duality of structuration. Rather is the problem a more general one. Society has to cope with structures that are not constituted through social practice. To reduce the agenda of human geography to that of social theory serves to gloss over these suppressed if not hidden structures. This is not to deny the value of concepts like Soja's spatiality or Giddens's structuration, but to argue that they can only form a subset of a much wider agenda concerned with the interaction between human agency and its conditional structures. This wider agenda needs in the first instance to recover human geography's concern with the predicament of humankind in its totality. With this in mind, I want to suggest that we need to see the interaction between agency and structure as articulated at different levels of the human condition. At its most fundamental, there are the manifold problems posed by how human agency interacts with its most deeply rooted cognitive structures in a socio-biological sense. More familiar, is how they interact at a cultural level, or through so-called traditional forms of behaviour and cognition that are deep rooted but learned and communicated inter-generationally. At risk of having it condemned as an outworn distinction, one no longer valid given the recent acculturation of social theory, I would further distinguish this cultural level from the social. As a working definition, I see the latter as comprising the norms, rules, practices and relations developed around the goals or projects that are operational within a society. In character, these are communicated or instilled through interaction with one's contemporaries or peers so that, in essence, they constitute an intra-generational form of knowledge. Beyond, or at a still higher scale of resolution, I would place the socio-organizational level of interaction. I see this as the level at which society is constituted as a functioning system, with roles, codes and functions. Finally, there is the socio-technical level of interaction centring around society's interaction with the physical or technical means through which it organizes and sustains itself as a system. It is at this level that we can consider the interaction between human agency and a society's built environment, the technical means by which it produces its material and energy needs and by which it stores and handles information, and the means by which it deploys its physical and technical resources of social power.

In reality, of course, these different dimensions overlap and intersect in a more seamless way. Indeed, a reading like Lefebvre's would reduce them all to a more or less continuous flow of social interaction, both process and product. However, their distinction, if only for the purposes of definition, has this advantage. I want to suggest that the problem of how human agency and structure interacts varies at each level. Any debate over how they interact, therefore, has to be ordered around them. To put my point more bluntly, what we label and treat as structure will have a more autonomous or deterministic role when seen at the socio-biological and socio-cultural levels, even at the socio-organizational and socio-technical levels, than when seen at the purely social. For the human geographer, the problem with structuration theory is that it is conceptualized in such a way as to handle the debate only at the social level, or the level at which structure can reasonably be described as embodied within rather than without everyday social practice.

When we consider other levels, the case for a more autonomous or determinate form of structure becomes much stronger. Social practice, for instance, can hardly be said to reproduce deeply rooted cognitive structures. Such structures are too hard-programmed. Not only are they not open to easy revision, but also, they are reproduced and transformed on a timescale that goes beyond what is meaningfully encompassed and reconstituted through the medium of everyday social practice. Though the findings of socio-biology have not found favour amongst social theorists because of the inevitable demarcation disputes over where explanation lies, such as in the nurture–nature dispute, they undoubtedly have significance for the spatial reasoning of human cultures. Given the emphasis which some geographers now attach to language and to the role of intersubjectivity, it is worth recalling how a linguistic philosopher like Chomsky has endorsed the structuralist view that language forms part of 'a general system of cognitive capacity that is determined by the innate faculties of mind' (Chomsky 1976: 123). For Chomsky, language, like 'our concept of physical space and the objects within it', probably conforms to a general cognitive system which 'set limits on human intellectual achievements' or 'objective bounds' without, he hastened to add, limiting its capacities (ibid.: 123–4). In fact, seeing the problem at its most fundamental level, no geographer interested in the cognitive basis of geographical order can ignore the work of socio-biologists on the two-sided or bicameral nature of the brain, with spatial and visual skills, together with emotional qualities, being located on the right side of the brain and analytical and mathematical skills on the left side. It is the right side that is 'important in perceiving spatial relationships' and in creating 'a coherent concept of the spatial organization of the outside world' (Nebes 1977: 104). Much is still to be understood about the bicameral structure or symmetry of the brain but even in terms of what is already understood, it is clear that humankind's ability to think spatially is a distinct skill and one that is functionally different from, say, analytical or mathematical skills. Dare one suggest that a minimalist view of Gregory's *Geographical Imaginations* (1994) – a text which strives, and does so brilliantly, further to shift our interpretative approach from an objective, mathematically based methodology to one capable of articulating more subjective and more richly nuanced themes – would be to say that it has succeeded in shifting thought within the subject of geography approximately four inches, that is, in making the bicameral leap from left- to right-hemisphere thinking (cf. Williams and Findlay 1991: 135–54).

Socio-biology has other conclusions whose deep structural significance makes purposiveness in human systems a relative matter. Recent work is especially interesting for anyone concerned with how culture relates to problems of space and environment. In stark terms, the consciousness which some, after Habermas, would have us put at the centre of geographical analysis is developed around a brain which some socio-biologists would see as a hunter-gatherer brain within a modern skull (Cosmides, Tooby and Barkow 1992: 5). The point being made is simple. Even considered only since the Pleistocene, humans have practised either some form of foraging or hunter-gathering for 99 per cent of their history. Given the slow rate at which brain architecture evolves, our complex brain functions, including our spatial sensitivities, are 'adapted to the way of life of Pleistocene hunter-gatherers' not to those conditions that we would commonly associate with human culture (i.e. agriculture, government, education, urbanism, etc.) (ibid.: 5). Furthermore, until comparatively recently, most human brain functions were domain specific, related to specific tasks in specific circumstances. Only around 40,000 years ago, at a time of considerable cultural change – with the emergence of language, symbolism, artistic representation and pair bonding – did humans develop what can be called domain-generalized skills, applying skills acquired in one domain to other domains. Amongst the oldest functions are those to do with survival: these include the basic foundation of our spatial skills. As an argument about more than 99 per cent of human history, it would require considerable faith in the capacities of human free will and the achievements of the Modernity Project, a process covering the last 0.05 per cent of human history, not to see this evolutionary context as having no structural import or carry over for modern culture. In the civilizing process, we may choose to ignore it completely but the question remains of whether we really have that choice.

At the other extreme, there is also a case for seeing the physical structures embodied in the built environment as hard-programmed because they too are not freely reproduced as a matter of routine. The once much-quoted saying that process creates structure but, once created, structure inhibits process is still a valid working hypothesis here. Patently, the built environment is socially produced, usually at a marginal rate, but there is no meaningful sense in which we can say it is continually or recursively reproduced through social practice with each instantiation. For this reason, it cannot be said invariably to map or express the prevailing needs of everyday social practice. Seen *in toto*, its form is invariably compound, both physically and symbolically. Societies create, add or replace it with each generation in a largely piecemeal, marginal way, though occasionally – such as with industrialization and urbanization in post-revolutionary Russia – large-scale new landscapes can be created with impressive speed. As a consequence, the prime quality of the built environment is its inertial rather than topical form, an inertia that develops from the very moment of its creation and serves to distance it from the social practices and processes that operate across it at any subsequent moment. We can deal with this inertia in two ways. Either we can ignore it, treating it as a source of noise that can only interfere with the development of a social theory grounded in social practice and those soft-programmed structures that are recycled through everyday routines, or we can try to incorporate it as a factor into our concepts through its potential role as a source of constraint. In fact, there is a case for arguing that much of the debate over the tension between fixed and circulating capital within the global economy has been inclined to treat the problem of fixed capital, its 'thingification', as a constraint despite Lefebvre's philosophical efforts to maintain thing and flow as part of a single process. The recurrent crises of capitalism, as Harvey has indicated, occur precisely because investments in the built environment, once made, are irrecoverable. The whole debate about uneven development, Harvey's 'spatial fix', and issues like flexible accumulation can be seen as part of a debate to conceptualize the problem of constraint in so far as it concerns capital. Likewise, some of those who have argued the case for structuration within human geography have, not surprisingly, struggled to keep structure from asserting its influence if not its autonomy. Thus, on the one hand, Soja talks about 'socially-created structure' (Soja 1989: 80) and the 'practical consciousness of the knowledgeable human agent' (Soja 1985: 120), but on the other, he also talks about 'the constraining and coercive tendencies arising from the enduring structural determinations of social life' (ibid.: 120) and, more critically, how spatiality was 'constrained by circumstances "directly encountered" in the already constituted spatiality of social life' (ibid.: 98).

The point which I am trying to establish here is that if we are to develop a sufficient understanding of societal change, and to draw out its underlying geography, then we need to recover an interest in those structures that inter-act with human agency in the broadest possible sense. Doing so introduces structures that have more autonomy or which manifest more constraint and boundedness in regard to social practice than can be contained within the framework of structuration theory and its concepts of duality and recursiveness. Indeed, far from bundling structural determination, constraint or boundedness off-stage, we should try to model its effects more explicitly in our thinking about change. To do this, we need an approach that is better able to define the circumstances in which structure, whether biologically given or socially and culturally produced, can affect everyday social practice in ways that are more or less non-recursive. Of course, to say that such structures are non-recursive is simply another way of saying that they have inertial properties, a tendency to resist easy change. In other words, any understanding of change must begin, or at least incorporate, a clearer understanding of the circumstances under which society does not change, either because it finds change difficult or positively resists it.

I want to argue that such an approach is intrinsically geographical because the way in which society reifies itself, turns concept to actuality and adapts to living in a material world, is a prime source of this inertia. Far from being a 'vacuous' concept, adaptation, society's commitment to a particular form, at least when meant in terms of how society is spatially disposed or organized and how it symbolizes or values its world, is capable of yielding rich insights into inertia. It is also a means by which we can negotiate a different sort of relationship between structure and human agency. This is because once we see inertia as geographically emplaced, then it introduces the possibility of human agency and structure interacting not just temporally, through phases of recursive or dialectical succession, but through strategies of spatial negotiation, with areas in which the emergence of new practices and new institutional forms takes place relatively easily being juxtaposed with areas in which it is prevented, retarded or constrained by structural inertia. The challenge of any geographical contribution to a theory of societal change is to expose this strategic use of space.

Social theory and the geography of societal change

Far from being a novel suggestion, I would argue that this attempt to see societal change as arising out of the structural constraints and possibilities that are mapped into the geography of a society is already a part of the debate over change within social theory, either implicitly or explicitly.

Giddens, for instance, talks about change occurring at the 'edges' between societies though nowhere is he really clear about precisely what he means by this in processual terms. Yes, he depicts edges as the points of contradiction and conflict between systems, but his stress on change being contingent suggests that he does not see any structural role for edges, though his argument clearly lends itself to one in which such a geography of structure has a place. As Gregory observed, he has 'so far not shown how the transformative capacities of human beings vary according to the specific circumstances in which they find themselves' (Gregory 1989: 379). By comparison, Mann's observations on change are more explicit about such circumstances. His analysis of early patterns of socio-cultural change emphasizes the way it unfolds at the geographical margins of society, erupting in the periphery of established systems where it is 'less bound by institutionalized, antithetical power structures' and shifting the loci of systems through time and space (Mann 1986: 539). He does not conceptualize it as a theory of societal change. Yet more so than in Giddens's argument, he seems to allow for the possibility that such a pattern of change is as much about inertia and structural constraint as about the seizing of new possibilities, peripheral areas displaying greater propensity for radical innovations.

A more explicit conceptualization of this idea is provided by those who talk openly about the structural diversity present within societies, with different parts manifesting different degrees of responsiveness to change. Such an idea is explored by Unger in his characterization of social institutions in terms of their entrenchment or plasticity. It received a more geographical treatment from Turner, with his notion of liminality and how edges or spaces of transition open up new possibilities (Turner 1974: 231–71). These marginal spaces create a social world of 'betwixt-and-between' (ibid.: 232). They fuse ideas and values from different societies, having 'a marginality of mind' that, for Turner, has often produced 'originality' in writing, art and philosophy (ibid.: 23). In recent years, such ideas have matured into a cultural theory of marginal spaces. As part of his project to question the idea of cultures as having a stable, stereotypical or 'originary' form, Bhabha portrayed such spaces as a 'third space' (Bhabha 1990: 211), an 'interstitial passage between fixed identifications' which 'opens the possibility of a cultural hybridity' (Bhabha 1994: 4). For hooks too, the subjection or dependency that we normally associate with such peripheralized, marginal spaces can be inverted so as to turn them into a 'site of creativity and power', one that possesses the power and freedom from constraint 'to imagine alternatives, new worlds' (hooks 1991: 150). No doubt informed by the thinking of writers like hooks, Hall has drawn on such ideas in similarly arguing that the 'emergence of new subjects, new genders, new ethnicities, new regions, new communities' have acquired a voice for themselves 'sometimes in very marginalized ways' (Hall 1991: 34).

Given their strong spatial component, it is not surprising that these and similar ideas have also been taken up and developed to effect by geographers like Shields and Soja. The focus of Shields's study was the geography of peripheral communities, or what he variously termed 'marginal places', 'geographic marginality' or the 'social periphery' (Shields 1991: 3). Far from being empty categorizations of space, their contrast with core areas, their sense of being 'the Other', permeated their whole identity. Shields refers to a culture of the periphery, one with its own complex landscape or web of schisms, interstices and boundaries (ibid.: 60). Though his approach is indebted to Lefebvre (ibid.: 52), with process and product being made to shade into each other, he also uses Bourdieu's ideas. In particular, he makes uses of Bourdieu's notion of habitus as a durably installed set of structuring principles, 'an invention within limits' (ibid.: 34) but, presumably in deference to Lefebvre, questions Bourdieu's 'redundant' use of the term 'structured structures' because they may be 'always both structured and structuring', or simultaneously product and process (ibid.: 33). In fact, Bourdieu's twin-notions of field and habitus can be made to lean the other way. Each society is seen by Bourdieu as comprising a set of fields. These fields form relatively independent areas of social activity or 'play', each with its own norms and principles of operation or habitus. As Wacquant has put it in an analysis of Bourdieu's work, these principles 'delimit a socially structured space in which agents struggle, depending on the position they occupy in that space, either to change or to preserve its boundaries and form' (Wacquant 1992: 17). The notion of habitus is a 'matrix of perceptions, appreciations and actions', a form of structure that exists within society to shape its response to situations. In a sense, the idea of habitus lies midway between the idea of society structured by factors outside of its immediate self and the idea of structuration with its stress on structure as something embodied in practice. It concedes the influence of structure as something historically and institutionally constituted (cf. Wacquant 1992: 19), but allows human agents some degree of freedom to operate *within* these structures. In the final analysis though, habitus is presented as a collective form that operates and changes beyond the individual, even at superorganic or transcendental level (Bourdieu 1992: 189).

Soja's recent review of third space also blends ideas from Lefebvre with those on marginal spaces being developed by cultural theorists like hooks. In what is the most sustained analysis of third space to date, Soja sees it as a melting pot where everything comes together and all oppositions stand face-to-face (Soja 1996: 56–7). Yet arguably, he emphasizes third space's imaginary rather than real-world position as a peripheral space. This comes across from his reading and reinterpretation of how Lefebvre's own background provided him with a third space within which to develop his own philosophical hybridity. Soja juxtaposes Lefebvre's peripheral origins in the French Pyrenees with his academic life in Paris. However, it is not the different perspective of the former per se that enabled Lefebvre to think ideas anew within the latter, despite what Soja calls his 'deeply peripheral consciousness' (ibid.: 30), but the interaction between the two, his life betwixt and between. As Soja himself says, this 'thirding-as-Othering' takes the problem beyond simple oppositions like those based on core–periphery or marginal spaces (ibid.: 60) but, in the process, it might be argued that it loses some of its essential spatiality for all the use of the term third space.

Indeed, some would argue that we do not need such an abstract reading to root change in its geographical context. Even a conventional reading of society in core–periphery terms can provide us with the basis for a contextual theory of societal change. Since Wallerstein first outlined his world-system theory, the germ of his idea has been developed in two ways. First, more and more scholars have argued the case for the existence of pre-capitalist world systems in which dominant cores exploit dependent peripheries through ties of political, social and economic hegemony. In other words, far from being the child of modern capitalism, the emergence of core–periphery systems, with all their implied structural and processual differences, are seen as part of the constitutive character of all societies and as shaped through a range of different types of social and political processes. Shils's distinction between core and peripheral communities and his purely sociological analysis of their differing propensities for upholding societal norms or changing them, one that can be aligned with the work of hooks *et al.*, provides us with the sort of dynamics of change needed to drive these pre-capitalist world systems (Shils 1975: 3–4). Second, just as some social theorists were demonstrating the universality of core–periphery structures and processes, pre-capitalist no less than capitalist, others have been refining how such forms of uneven development form an intrinsic part of a capitalist society, with capital continually seeking out differences in labour and resource cost, and disinvesting as soon as these differences threaten to become equalized (Smith 1984; Smith 1986: 87–104). In doing so, it provides the dynamic for the production and destruction of space as a 'crude and restless auteur. . . never fully able to make up its mind' (Soja 1989: 157). Indeed, its as yet uncharted history forms the basis for Harvey's project on the historical geography of capitalism (Harvey 1989).

Though it covers only part of the argument which I want to develop, the work of these scholars enables me to locate a starting point for my argument firmly within the debate over social theory. Societies are developed around diverse rather than uniform structures, their structural form varying both between different types of space and in how they relate to everyday social practice. These structures release, constrain or deflect the pressure for change. Already running through a range of socio-theoretic argument is the assumption that in all societies or systems, a fundamental contrast exists between those parts that are core and stable and those parts – the marginal or peripheral spaces – that are not. In character, they can be cross-matched against Unger's areas of a society marked by conditions of institutional entrenchment and those marked by conditions of institutional plasticity. One is potentially context preserving and the other, context changing. Clearly, the challenge for the geographer is to explore how such contrasts are constituted and mapped and to elucidate how they might provide the basis for an interpretation of change. However, compared to recent debate on change, I want to build an argument that assigns relatively more importance to structure and one that recovers and redefines the role of inherited or past forms in the process of change.

I want to take the argument forward in five broad stages. First, in the following chapter, I want to contextualize my discussion by looking at the wider debate over change, drawing out how geographical perspectives currently relate to it. Second, in chapters 3–4, I propose to explore the actual record of societal change as a geographically framed experience. In particular, I want to show that simple assumptions about the sustained, ongoing nature of change carried forward by societies performing as complex, continually adaptive systems or by 'reflexive beings' continually mastering their own destiny have only limited support in the actual record of change. We can demonstrate the point at both the global level in regard to empires and world systems and at the level of the individual society in regard to the state and its regional parts. Third, in chapters 5–7, I want to lay the foundations for an understanding of why change ultimately appears to seek out new societies and spaces by exploring the different ways in which geographically emplaced forms of structure become inertial, constraining or deflecting change. Altogether, I want to consider the inertia which is developed through symbolical, organizational and fixed-capital forms. Fourth, on the basis of this systematic discussion of inertia, I want in chapter 8 to consider the extent to which we can conceptualize the way in which societies deal with their inherited past and whether this provides us with a more geographically sensitive interpretation of how human agency and structure interact. Fifth, in a concluding chapter, I want to explore how inertia prepares the way for a geographically informed concept of change, one that sees it as exploiting the difference between areas of freedom and unfreedom, flexibility and constraint, in a strategic way.

2

Establishing a taxonomy of societal change

Intellectual knowledge, like vision, is fragmenting: we need apartness. We come to tell what one thing is by cutting it off from other things – and 'whatness' is of the essence of our knowing. We know by putting together what we have taken apart
                                                                        (Ong 1977: 138)

By way of an introduction to a geographical perspective, I want to suggest that we need to establish a more systematic understanding of the wider debate over change. Despite the vast amount that has been written, there have been surprisingly few attempts to define what constitutes a sufficient or rounded interpretation of societal change. Arguably, it is a more complex and multi-faceted process than some interpretations allow. Indeed, many disagreements in the literature can be explained away by the simple fact that supposedly rival views are addressing different aspects of the problem. We can best compre-hend these differences by trying to construct a taxonomy of change, one that, first and foremost, identifies the different dimensions through which it needs to be analysed, and which then orders actual viewpoints around these different dimensions (see Tables 2.1–4). Through such a taxonomy, I want to draw out the fact that there is already a basis for establishing a geographical perspective on the debate. The task facing the geographer is to give this component greater definition and to set about conceptualizing it.

In theory, a taxonomy of societal change should distinguish between, on the one hand, the nature of society as an ongoing system and, on the other, the dynamics that bring about change. In practice, such a neat and tidy ordering of the debate is not possible. For some, society is organized for stability so that change appears as a dysfunction, something that society does not consciously or intentionally seek. It arises as an externally driven process, involving dis-tinct causes and means. Others, meanwhile, work from the assumption that society is organized for change, indeed, that it positively seeks change. Because it arises from internal factors that are part and parcel of the way society func-tions normally, it becomes superfluous to differentiate the dynamics of society as an ongoing system from the dynamics of change. Occupying a midway position is a further line of argument which stresses the extent to which change is something that goes on between societies, around their edges or at their margins. Though it would be misleading to impose too much uniformity on this particular class of ideas, they have in common a belief that change is not reducible to either external causes or to normal societal processes but is an epiphenomenon that arises out of special circumstances, opportunities and conjunctions that occur around the edge of societies.

Clearly, if we are to cope with such differences, we need a taxonomic matrix which is able to distinguish the *dynamics* of society as an ongoing system from the *causes and mechanisms* that bring about change, yet which acknowledges that when seen in the context of some interpretations, the dynamics of society and the causes of change are conflated into a single dimension of analysis. The question of what change amounts to, its *sum product* as a process, provides us with a third dimension of the problem. Though outwardly a straightforward even descriptive dimension, this is an aspect that embraces the most profound disagreements over whether change has a direction, teleological or otherwise, and whether it is progressive or evolutionary in any sense. We can distinguish a fourth and final dimension based on the debate that exists over what can be termed the *morphology* or *trend* of change.

The dynamics of societal systems

The dynamics of societal systems can be grouped around three basic interpretations: those which stress the systemic, structural or organizational characteristics of society; those which stress the paramount role played by human agency; and those which stress the interaction between structure and agency (see Table 2.1).

Systemic approaches

Amongst systemic approaches, the simplest is that founded on the notion of society as a simple equilibrium system. This provides us with a rigid, mechanical view of society, one that sees it as unable to cope with any sort of fluctuation outside narrow limits of tolerance. Stated crudely, it seems to offer a cynical view of society, one that denies it any sort of capacity for steerage. Somewhat more sophisticated are those interpretations which treat society as a homeostatic or cybernetic system, conceding it the means to dampen flux around its base conditions via various forms of negative feedback. It still amounts to what we can call a 'one-shot' system organized around fixed goals, but it now has the means to regulate its behaviour, to self-repair, so as to maintain these goals better in the face of fluctuations. In the literature, homeostatic systems are aligned with structural-functionalist views. By stressing the struc-

Table 2.1 The dynamics of societal systems

	1	2	3	4	5	6	7
A. Systemic/structural approaches							
	1a. Mechanical, single state systems, non-adaptive to change	2a. Goal seeking-order maintained through buffering and use of regulators to dampen flux and deviation	3a. Self-maintaining and -organizing system – metastable order maintained via buffering and use of regulators but capable of switching to new adaptations when need arises	4a. Self-organizing and -steering system – order maintained via both regulators and controllers and able to switch from system-conserving to system-changing	5a. Self-steering system – ongoing change within order achieved through regulators and controllers, with latter dominating	6a. Progressively adapting system – achieved through differentiation, specialization, linearization, hierarchization	7a. System affected by system needs – social process periodically driven across non-linearities and bifurcation points
	1b. Simple equilibrium systems – change in state conditions = crisis/collapse	2b Homeostatic/cybernetic models – order maintained via negative feedback only	3b. Society as adaptive system – metastable order maintained via negative feedback and occasional use of positive feedback in crises	4b Society as complex adaptive system – shifts regularly between self-organizing and self-steering system	5b. Society as complex adaptive system – change as continuous, ongoing process	6b. Society as hyper-adapted system – changing continually by adaptive upgrading	7b. Society as non-linear system – state conditions shaped through system needs and explained through catastrophe theory, chaos theory, non-equilibrium systems
B. Human Agency							
	1. Action-orientated and voluntaristic models	2. Behavioural, role of cognition, perception	3. Primacy of class conflict models	4. Analysis of society through discipline of context, all processes seen as contingent, conjunctural or situated and anchored to specific contexts			
C. Structure–agency interaction							
	1. Structuration – structure and agency part of duality – interact recursively, open up structure to everyday social practice and re-interpretation, error copies	2. Structuration – structure and agency interact recursively – everyday social practice and new experience interpreted via a priori concepts, limited potential of primary tropes	3. Communicative rationality – based on rationalized life-world ordered around structures of linguistically produced intersubjectivity and how they interact with events, new technical demands and capacities	4. Role of formative or imaginative contexts and dis-entrenchment – interplay between context-transforming and context-preserving situations – push of sequential effects and pull of plasticity	5. Interaction between events, conjunctures and structures – differences in scale and chronological phasing – differences in the immediacy and randomness of events, the cycles and trends of conjunctures and the relatively stable slow-moving forms of structures		

ture- and boundary-maintaining processes of society, and their complex inter-dependency of roles and parts, structural-functionalist thinking inevitably depicts society as organized for stability or stillness rather than change (Bock 1963: 232).

A fundamentally different family of systemic interpretations starts from the assumption that society is a dynamic rather than stable system, with the capacity to switch between different states. In addition to the use of negative feedback to dampen fluctuations deemed threatening, there is now positive feedback to respond to new opportunities and possibilities. Framed in such terms, societies become open and negentropic, able to absorb new information and to elaborate their structure in response to new opportunities and possibilities. Two sorts of systemic interpretation are usually invoked here. First, there are those which combine society's capacity for making fresh adaptations with the notion of metastability. In effect, such an approach negotiates a compromise between equilibrium and dynamic systems. It assumes that societies work to establish durably stable equilibrium states, but that, over time, such equilibrium or metastable states are stressed by external threats such as environmental crises. However, being essentially open systems, societies are able to respond to such crises by reorganizing themselves around new metastable states (Butzer 1980: 517–23).

A second group of systemic interpretations works from the assumption that continuous, ongoing adjustment between states is part of the very nature of society. Altogether, we need to differentiate between three sub-types. First, there are those which treat societal systems as self-steering or purposeful systems because they possess a high degree of steerage. Invariably, such systems are linked to 'progressive' models of societal change, with society constantly looking for self-improvement through more elaborate adaptations, mapping more of the variety of their environment into their organization and evolving more complex control hierarchies. In its most developed form, this type is exemplified for us by Buckley's notion of society as a complex adaptive system, one that is constantly adjusting its state in response to threats and opportunities (Buckley 1968b). Second, there are what can be called self-organizing systems. These stress the extent to which human systems function as self-repairing systems without necessarily being purposeful. Their organization experiences irregular moments of readjustment in response to random, contingent events but without being guided by specific over-arching goals. In a phrase, they seek to persist in a given form but periodically have to rethink how they do so. Where formally stated, such models usually make use of concepts drawn from chaos theory, non-equilibrium systems or catastrophe theory (e.g. Allen 1988b: 101–4). Finally, a third type of systemic interpretation rests on the assumption that the nature of societal systems is best understood by modelling the constraints acting on them rather than their internal dynamics. Behind such an approach is the assumption that societal systems are either too complex or too unpredictable to be modelled in any direct way and that, in consequence, we gain more by trying to understand what may constrain their range of possible states rather than by trying to model what determines any particular outcome (Gould 1972: 689–700)

Society and human agency

Providing a counter-view to overtly systemic interpretations of societal systems are those based on the absolute primacy of human agency. In their most extreme form, these strip all systemic or structural imperatives away from the problem. Instead of a large-scale, over-arching perspective on societal forms, the view from the perspective of angels (Harvey 1989: 317), the problem is reduced to human needs, norms, values, perceptions, intentions, actions, events and situations or locales. Though allowed context, association, contingency and succession, the impact of such factors is read in a non-deterministic or non-structured way. Each community and its experience of time and space is treated as individual to itself and as capable of only being understood as such. In contrast, there are other interpretations which, whilst reducing society down to the level of individuals or groups and the paramountcy of human agency, still prefer to conceptualize their reading of society, building their interpretations around voluntaristic, action-oriented, norm-oriented, purposive-rational, behavioural or class-conflict models. Werlen's use of an action-oriented approach to redefine human geography, with its explicit adaptation of purposive-rational models and their economic derivative, rational choice theory, captures the essence of such interpretations. It is an approach which, first and foremost, is ego-oriented or based on an I–we perspective, and which, in principle, starts from the subjective or inter-subjective meaning-context in which all actions are rooted (Werlen 1993: 100–38 and 179–206).

Over the past decade or so, theories of society built around the primacy of human agency have taken on board a new set of themes that stress issues of cultural identity, image and representation. Acting as context for this acculturation of social theory has been the increasing domination of all communities and spaces by the global economy and the imperatives of capital since the Second World War. Globalization is implicated in a revised social theory in two ways. First, it has overridden indigenous cultures and their identity, smothering them with the values, roles and routines of a wider system. Apparently homogenized systems of this sort have existed before. The so-called axial systems of the medieval period imposed a uniformity of authority and belief over large areas, but their uniformity amounted to no more than what Gellner saw as a 'capstone' set over a welter of localized cultures and practices (Gellner 1983: 9). What is different about the growth of the world system in recent decades is the way in which it has insinuated itself downwards into everyday social practices, changing their constitutive character in fundamental ways. The result has been the erosion of traditional local cultures and their replacement by new roles and identities imposed by the global economy (Wallerstein 1991: 184–230; Hall 1991: 19–39). Yet even in the most advanced society, this process of penetration is still only partial, so that local societies consist of a bedrock of indigenous or local culture overlain by an invasive and expansive global culture. Inevitably, the juxtaposition and interaction of the two has fostered crises of local identity and a growing concern with questions of image, representation and difference (Featherstone 1995: 6–14 and 86–101).

The second way in which globalization has led to a revised social theory, one invigorated by a heightened concern for culture and ethnography, is through the instability brought about by the mobility of capital. In those regions penetrated by it, capitalism can create communities *de novo* through major phases of investment. In time, as returns on investment decline, capital is disengaged via interest and returns on investment and redirected to new regions and new technologies. In this way, communities eventually find themselves with a stock of fixed capital locked into yesterday's production functions. More serious, communities which had evolved around them and which had attuned their cultural practice to particular locales are left with a crisis of disinvestment. This cycle of growth and decay poses new problems of social identity, problems which focus attention on the way capital produces and discards places.

More recently, some have taken this argument further. They have responded to the disorientation and dislocation engendered by global capital by stressing the need to comprehend society through its constituent sub-groups and communities, each culturally constituted in its own terms and each to be studied as such. Indeed, some would now proclaim the emptiness of concepts that deal with society as an integrated or totalized form. In Lyotard's blunt words, we should 'wage war on totality' (Lyotard 1984: 82), though much of the force of his criticism stemmed from the fact that he equated the 'conception of society as a totality' with totalitarianism (Readings 1991: 111). For these post-modernist approaches, all forms of metatheory or large-scale narratives are invalid, as are the assumptions fostered by the Enlightenment Project that there is such a thing as an objectively rational and universal concept of society waiting to be discovered. Once the notion of an objectively rational basis for interpreting society is abandoned, as well as the notion that there are forms of knowledge or culture that are self-referential, then all knowledge about society and its organization becomes relative, constituted around the norms envisioned by particular cultures and social groups. Stripped of a unifying concept, society is no longer the sum of its parts, but an assemblage of values, world-views and locales that need to be considered as such. What now matters is how the parts differ, their sources of identity as expressed through social practice, image and ethnography.

In its most uncompromising form, such interpretations stress the disjunctural nature of time and space for contemporary society, with the past being reduced to disconnected images and, thanks to the disorientating effect of global communications, the human body losing the capacity 'to map its position in a mappable external world' (Jameson 1991: 44). Yet when we probe further, we find that the case against an integrated or totalizing view of society often rests more on anxiety about how it contributes to a potential theory of power and exploitation than with its failure to describe the world authentically (e.g. ibid.: 331–3). In fact, the post-modernist concern for a concept of society rooted in issues of local identity and representation may stand in opposition to the metatheories sustained by globalization but it hardly does so at the expense of such theories. A measure of this is provided by the extent to which studies in the cultural politics of place see local spaces as an arena of competition between the local and global. Just as local communities increasingly use space to assert their separate identity, and increasingly respond to globalization via diverse strategies of local governance (Goodwin and Painter 1996: 635–48), so also does the global economy use local space to symbolize its power and presence (Ley 1987: 40–56; Zukin 1991: 25–38).

The interaction between agency and structure

Providing a third class of interpretations are those which combine the attributes of a systemic approach with attributes offered by a voluntaristic or action-oriented approach. Avowed system theorists like Buckley and Isard made elaborate attempts to move away from systemically determinate models of society by trying to incorporate a greater sensitivity to the human agents and groups whose lives are modelled by such systems, but they did so only by trying to factorize it, adding agency as another quantifiable input into their models (Isard 1977; Buckley 1968b). Others have formulated views that balance systemic ideas against the voluntaristic or action-oriented attributes of society in more equal measure. Parsons, for instance, saw society in systemic terms, but he constituted it as a system based on dimensions of action rather than control, distinguishing between four sub-systems of action: the pattern maintenance provided by the cultural system, the integration provided by the social system, the goal attainment provided by the personality system and adaptation provided by the behavioural organism (Parsons 1966: 28–9).

To a degree, Habermas's concept of society follows Parsons in trying to link an action-theoretical approach with a systems-theoretical approach, yet this point made, their views differ sharply on one fundamental point. Whereas Parsons conceptualized the lifeworld within systems-theoretical concepts, so that 'the structural components of the lifeworld become sub-systems of a general system of action' (Habermas 1987: 153), Habermas looked to invert their relationship. His theory of society accords a more fundamental role to the rationality structures of the lifeworld 'that find expression in world views, moral representations, and identity formations, that become practically effective in social movements and are finally embodied in institutional systems' (Habermas 1979: 98). Admittedly, there are times when he sees the 'competition' between action-theoretical and systems-theoretical paradigms as being as much to do with differences of perspective as with any hierarchy of determination (ibid.: 105). However, in his work as a whole, he leaves no doubt that he sees the irreducible base for any analysis, 'the horizon behind which we cannot go' (Habermas 1987: 149), as formed by the lifeworld of societies. He builds an understanding that works outwards from this intersubjectively shared lifeworld and from the intercommunicative understanding through which societies reproduce the structures of their rationalized lifeworld. For Habermas, our grasp of this lifeworld and the intercommunicative understanding which it sustains cannot be reached through an observer-based system perspective for the systems developed around society are ultimately determined by the structures of its lifeworld not vice-versa (ibid.: 153). From an analytical perspective, the problem we face is that the concept of the lifeworld is accessed by members of a society 'intuitively' (ibid.: 136). It is the 'vast and incalculable web of presuppositions that have to be satisfied if an actual utterance is to be meaningful' and if members of a group are to reach a mutual understanding through such utterances (ibid.: 136). Whilst Habermas talks about such a lifeworld as a concept, or a system, as something rationalized and as having structure (i.e. Habermas 1984: 43), he would also claim that it cannot be objectively measured or analysed in such terms.

Giddens's structuration theory provides an important third variant in this debate over whether concepts of society should stress human agency, or action-oriented approaches, as opposed to the system-theoretic qualities of social systems. Like other theories based on the notion of duality (Thrift 1996: 67–72), it does so not by trying to fuse them or by negotiating away their differences, but by transforming their perceived dualism into a recursive relationship, a 'duality of structure'. The systemic and organizational structures of society are seen as sustained only through everyday social practices and vice versa. Each is, in effect, both the medium and the outcome of the other. Yet despite this interconnectedness, they are not seen by Giddens as of equal weighting in any concept of society. Systemic approaches create what he sees as a false agenda of system or functional needs. The more these systemic needs are asserted, the more it reduces society to a world of 'mechanical dopes' borne along by impersonal forces rather like Rupert Brooke's 'poor straws' briefly brought together on the 'dark flood', socialized, then carried away 'into the night apart'. Once we see social structures as realized through everyday social practices, though, they now become open to the will and action of knowledgeable, reflexive beings. Far from being beyond revision, they now become open to change through the unintended as well as the intended consequences of human actions.

We might compare these ideas with those that maintain the separateness of structure from agency. Berger and Luckmann, for instance, considered the institutional form of society to be 'experienced as an objective reality. It has a history that antedates the individual's birth and is not accessible to his biographical recollection', the individual's life being but 'an episode located within the objective history of the society' (Berger and Luckmann 1966: 61). Without destroying or glossing over the categories of structure and agency, Giddens's concept of structuration makes each accessible to the other, but one can hardly say that it preserves all their qualities. To squeeze structure through the medium of everyday social practice, structuration reduces structure to its pure socialized form. Other forms of structure, including forms of interest to the geographer, such as the physical structure (including urban forms, manufacturing systems, transport and communication networks, etc.) that underpins much of the spatiality of modern social life, are excluded.

Giddens's structuration theory presents the problem in terms of structure and agency, or structure and social practice, and how they can be brought together within the framework of his 'duality of structure' without losing their separateness, but as Sahlins especially has argued, we can repackage their dichotomy or opposition as the 'false' opposition between stability and change. Just as the distinction between structure and event is seen by him as 'pernicious', so also can it be argued that there is no ground for supposing an absolute distinction between stability and change, or past and present (Sahlins 1987: 153). However, for Sahlins, social action is the 'appropriation of events in terms of *a priori* concepts' (ibid.: 153). As with structuration theory, Sahlins acknowledges that groups may shift or even reinterpret concepts through their very act of applying them, what he calls the 'risk of the categories in action' (ibid.: 153). Yet whilst he accepts that each application of a cultural scheme may transform it, he crucially claims that 'culture is precisely the organization of the present in terms of the past' (ibid.: 155).

Sources of societal change

It will already be apparent from what has been said that the different theories of society outlined in the preceding section cope with the problem of change in different ways. Some stress how society is organized for stability so that external factors have to be invoked to explain change, whilst others see it as part of the normal inner functioning of social process. With this in mind, when we turn to the problem of classifying different theories of change, we need to draw a broad distinction between, on the one hand, those which define society in such a way as to exclude change so that it becomes something intrusive and disruptive and, on the other, those which see it as an integral part of the way society is actually constituted. By way of a midway position though, we need to allow also for those who see change as neither wholly external nor internal, but as peripheral or interstitial, a process that occurs around the edges of society. Once seen in these terms, then the primary ordering of how change originates can be handled under the three headings of external, internal and peripheral sources of change (see Table 2.2).

External sources of change

Social change and maladaptation

The social theorist, Nisbet, has argued that 'nothing is more obvious than the manifest tendency of society to resist change' (Nisbet 1969: 270–1). Such a view presupposes that society functions either as a simple equilibrium system or as a more developed homeostatic system, both of which would have the effect of locking it into a conservative or inertial form. This fixed equilibrium or homeostatic state can be seen as stemming from society's unswerving commitment to a particular hierarchy of control or resource distribution, or from its organization around a particular normative code or value-orientation. In effect, society adapts itself to the imperatives of a moral or ideological landscape, one that is built in the mind before it is mapped in reality (cf. Wheatley 1969). For others, its fixity has to do with society's adaptation to a given environment, either to the opportunities offered by its physical environment or those of its wider socio-economic and political context. As a concept, this sort of adaptation can be formally expressed as the extent to which a society maps the information or variety present in its external environment into its internal organization (Ashby 1968b: 135–6; see also, critique by Tilley 1981: 365).

Arguably, we might expect socially and technologically more complex societies to map greater amounts of variety into their organization than simpler ones, and to buffer themselves more effectively against external or environmental perturbations. In other words, we might expect them to use their more complex organization and technology to make adaptation mean less or to threaten less. In fact, we need to handle this point more carefully. By mapping more variety into their organization, complex societies are effectively making a greater commitment to a particular adaptation. The extent to which modern societies use organization and technology to buffer themselves against environmental perturbations may make this relative inflexibility appear less obvious but it is no less meaningful. By comparison, less complex societies develop more obvious, flatter forms of adaptation but, as Denevan has demonstrated, their adaptations are often more flexible (Denevan 1983: 401–4). In other words, far from leaving the need for an adaptational response behind in their development, complex societies might actually be more committed to them.

Whatever the basis for this adaptation, it has this consequence. Change becomes something that society does not willingly or intentionally seek. It

Table 2.2 *Sources of societal change*

	1	2	3	4	5	6	7
A. Externally driven	1. External – determining inputs from physical environment, leading to crisis and adjustment	2. External – determining inputs from socio-political and/or economic environment, including world systems models	3. External – response to opportunities provided by physical environment, due to environmental change or new technologies	4. External – response to opportunities provided by wider socio-political and economic environment	5. External – natural selection – role of macroscopic diversity, hill-climbers and stochasts	6. Society as complex adaptive system – change as continuous ongoing process – adaptive upgrading	7. Social change as non-linear process – state conditions shaped through system needs and explained through catastrophe theory, chaos theory, non-equilibrium systems and dissipative structures
B. Internally driven	1. Internal – immanence models – developmental, realization of latent potential, epi-genetic versions	2. Internal – immanence models – dialectical, change as resolution of conflict	3. Internal – evolutionary, involuntary – ontogenesis	4. Society as adaptive system – metastable order maintained via negative feedback and occasional use of positive feedback in crises	5. Society as complex adaptive system – shift regularly between self-organizing and self-steering system		
C. Edge processes	1. Inter-societal conjunctions and collisions – role of societal edges and interstitial surprises	2. Role of societal edges as areas of unused freedom and bearers of interstitial surprises	3. Role of marginal groups and spaces – concepts of liminality	4. Change through geographical remapping – geographical law of geographical displacement, epi-genetically based core–periphery models			

arises as something intrusive, a threat or opportunity from outside. Once seen in this way, environmental factors, defined in the widest possible sense, provide the only logical source of change.

A restrained use of this sort of externally driven change starts from the position that society has a choice over how it adapts to its environment but, owing to sudden or unforeseen shifts in its physical environment (e.g. climatic change, volcanic activity), its adaptation turns to maladaptation, forcing change or collapse. A society's slow response to environmental change, or even its poor monitoring of environmental change, may be contributory factors here. The most explicit use of such an interpretation is that provided by Butzer. Reviewing the dynamics of early socio-cultural systems, he stressed their tendency towards metastable equilibrium. They constructed adaptations that stabilized their relation to environment over fairly long periods, buffering themselves against normal levels of flux by means of technology and the close monitoring of variability. However, in time, the chance concatenation of different threats such as disease outbreaks or natural disasters like floods precipitate system-threatening crises. In these circumstances, such societies are faced with the challenge of collapse or readapting to a wholly new metastable equilibrium. For Butzer, the essence of early socio-cultural systems was that whilst they maintained themselves at metastable equilibria, they were capable of adjusting to a new metastable state when need arose (Butzer 1980: 517–23).

World-System Theory

The idea that environmental fluxes can stress societies is only part of the problem. We also need to be mindful of how recent 'world-system' thinking, particularly that advanced by writers like Wallerstein, has made use of the wider social, economic and political environment within which societies are embedded. A key proposition of Wallerstein's 'world-system' theory is that a world system has emerged since the 'long sixteenth century' and that increasingly, the internal social, political and economic character of states is shaped by the role which they have come to play within this wider world system. In other words, once embedded within the world system, then the character of a society or state in terms of its economy, financial system, political character and conditions of labour control, together with the conditions under which these aspects of its character change, is driven by the wider needs of the world system (Wallerstein 1974; 1979). The roles are fixed, only the countries which play them change. In terms of the processes involved, this is a wholly different sort of 'environmental' determinism from that rooted in the physical environment, but taxonomically, there is a comparison to be drawn between the two, each treating change as something prompted from without. However, there is a necessary qualification. As Wallerstein himself put it, once a world system is established and countries are bound into it, there is only one history, that of the system as a whole, so that subsequent change becomes internal to the system.

Survival of the fittest

Providing a different perspective on the role of environment as a source of change are those interpretations which invoke some form of selection principle. Societies, communities and institutions are seen as working out adaptations to their environment. However, relations between a society and its environment are infinitely complex rather than simple. In any given environment, a society can evolve a diversity of potential adaptations. Though these adaptations are worked out intentionally and knowledgeably, it is assumed that no one can anticipate what is the best answer to cope with short-, medium- or long-term flux simply because there is too much complexity and uncertainty about how conditions will change. As Nelson and Winter put it in relation to economic organizations, the implications of using evolutionary theory are that 'choice sets are not given, and the consequences of any choice are unknown' (Nelson and Winter 1982: 276). Ultimately, the solution best adapted to conditions succeeds best and displaces less successful solutions. However, the important point to grasp is that what constitutes a successful or even just a viable solution is only determined through subsequent practice and selection. In other words, diversity provides a hedge against uncertainty. In a rapidly changing environment – physical, socio-political, etc. – those societies which preserve diversity, even at risk of performing sub-optimally, have the best chance of survival. Seen in these terms, change becomes a selection process, with non-viable solutions dying out and viable and best-fit solutions succeeding (Nelson and Winter 1982; Porter 1990).

Considered conceptually, the idea of societal systems being selected for viability within a given environment is made interesting by the effect of two key inputs. First, there is the point mentioned earlier that whilst complex systems monitor their environment more closely and buffer themselves against its perturbations through technology and functional differentiation, their commitment to specific forms of organization is potentially a greater source of inertia and maladaptation when faced with further changes in environment simply because of the greater costs and vested interest involved. Second, with more complex systems, the hyper-integration provided by developed exchange systems has created far more potential for such maladaptation (Rappaport 1977: 66). If we put these points together, they would suggest a world increasingly organized for stability and even stagnation rather than change. A way out of this cul-de-sac of reasoning is provided by Allen's work. Evolution, he argues, is not about the selection of optimal behaviour, but about the selection of species or sub-systems that can cope with ongoing change (Allen 1988a: 28; Allen 1988b: 105–10). Put another way, evolution depends on microscopic variability and randomness not uniformity or standardization. It relies on the generation of such variety through mutations and error, and through 'stochasts' who do not respond to available information (ibid.: 116). Most of these 'errors' will be loss-making and will die out but amongst them will be 'hill-climbers' who can perform at a higher level of functional efficiency when faced with change (Allen 1988b: 107). Thus, at any one stage, the capacity of a societal system to adapt and to go on adapting in response to ongoing changes in its environment will depend on its capacity to generate or tolerate microscopic variability or potential 'hill-climbers' in its population and institutions. In other words, a liberal society is more likely to foster ongoing adaptations and survival than an autocratic or dictatorial one, just as a decentralized and uncoupled organization which tolerates local solutions is more likely to survive than a highly centralized top-down organization. The reason why lies not in the absolute efficiency with which society can be co-ordinated to address current conditions but in its ability continually to generate the variety of forms from which new solutions can be selected to survive and compete under new or changed conditions. In short, change becomes an interaction between the degree of choice exercised by a society and the continuous selection for both viability and success exercised by changing environmental conditions.

Internal sources of change

The debate over internal sources of change embraces a wide range of different views, including views that are radically opposed to each other. Altogether, we can group them under three broad headings. First, there are those which stress the systemic nature of society and ascribe a primary role to factors like system needs, stresses, thresholds and organizational constraints in the conditioning of society for change. Second, there are those which conceive change in terms of immanence, that is, as something arising out of the innate character of social systems. As a group, these can be sub-divided into those using some form of life-cycle concept and those based on some form of dialectical conflict. Third, there are those which stress the action-oriented nature of change, with human agents and their actions bringing it about. Within this broad category, we need to distinguish three different sub-types. For some, change is a problem that needs to be analysed solely through human actions alone. For others, it is to be analysed through the interaction of human agency and structure but with the former always being the determinant of change. Still others have accepted the need to incorporate a greater role for human actions and intentions but have tried to do so by making systemic readings of the problem much more sensitive to such inputs via notions of society as a continually responsive system rather than one in which its systemic character fosters lags and inertia.

Society as system

The idea that the systemic nature of societal systems can affect their development through the operation of system capacities, constraints, needs and stresses is a well-founded theme in the literature, despite the criticism levelled at it by social theorists and human geographers. Underpinning it is the assumption that particular forms of societal or socio-cultural organization have an optimum level of operation. Obviously, it does not follow that a given organization of society is prevented from operating outside of this optimal range, but we can expect its efficiency and effectiveness to be diminished. Attempts to give such ideas a more secure analytical foundation have been made by various writers. Carneiro, for example, linked particular types of institution to the size of particular societies, demonstrating how the two were linked (Carneiro 1967: 234–43; Johnson 1982: 389–421). Tainter especially, has analysed how the economics of maintaining particular types of organization beyond their optimum range can collapse complex societies. Using an array of examples, he drew this conclusion. As a society becomes more complex, its maintenance costs also rise, so that an increasing proportion of energy has to be spent simply on maintaining its organizational forms (Tainter 1988: 92). In effect, societies are subject to a law of diminishing returns. Up to a point, they can successfully convert energy – as measured by investments in agriculture, crafts, education and so on – into extra socio-political organization. At first, the cost-benefits of such investments are favourable. However, sooner or later, a point is reached at which further units of energy investment do not yield a proportional increase in returns, but instead, yield smaller and smaller increments of return. For Tainter, such increases in the complexity of socio-political organization have an exponential effect on costs so that returns on further energy inputs are not commensurate. When societies reach this point, they are faced with the prospect of crisis and even collapse. His own conclusion, though, is that where a collapse back to simpler forms takes place, it can be seen as a rational solution to the problem of rising costs, enabling such societies to return to a more efficient level of energy use in relation to their socio-political organization (ibid.: 121).

Such ideas are echoed in work on institutional economics. Arrow, for instance, has stressed the huge irreversible investment attached to the formation of firms and organizations, largely through the learning of system-specific skills and codes (Arrow 1974: 55). These skills and codes introduce a potential for inertia since they 'are determined in accordance with the best expectations at the time of the firm's creation' (ibid.: 56; see also, Nelson and Winter 1982: 120 and 135). Olson has tried to conceptualize some of the reasons for this inertia. His central thesis is that where societies persist within stable boundaries, they tend to acquire coalitions and alliances, with the range and number of coalitions and alliances being related to how long the system *in toto* persists. Over time, the character of these coalitions and alliances alters. From being concerned with the expansion of output, they become concerned with how output is distributed. From being concerned with innovation and change, they become distributional coalitions concerned with maximizing their private gains from the system as it stands. The implications of such work for the debate over change are stark. The emergence of complex organizational forms has costs and rigidities that ultimately reduce their flexibility and capacity for change, so that any pressures for further change find it easier to work around them.

A wholly different way of applying systemic concepts to the problem of societal change is provided by those models which use concepts drawn from the physical or mathematical sciences, like catastrophe theory, maximum entropy modelling and non-equilibrium or chaos theory. When applied to societal change, they patently represent forms of interpretation that attach far more importance to structural or systemic factors than to human agency. In this sense, they come across as deterministic models, with system conditions determining the moment of change and even the new equilibrium state. Yet there is also a sense in which they combat the deterministic nature of conventional system models since they introduce a role for uncertainty. Considering them closely, catastrophe theory has been used both to explain change at a system and at a sub-system level. As a concept, it seeks to explain the non-linearities across which social change is sometimes driven. It is developed around the assumption that multiple equilibrium solutions can exist for certain types of basic system. At critical points, or close to what are called bifurcation points, the system can shift suddenly and catastrophically between these alternatives. Zeeman, one of the pioneers of catastrophe theory, has himself applied it to the dynamics of ideological change, demonstrating how the sudden jumps from authoritarian right-wing systems to authoritarian left-wing systems or vice-versa are really multiple solutions to the same basic system, with revolutions or coups representing the sudden switch between them (Zeeman 1979: 463–79). In a similar demonstration of its potential, Renfrew has shown how the notoriously unstable boundary between chiefdoms and proto-states can also be described using catastrophe theory. Within certain regions or phase states of a system, society can switch easily and rapidly between the two (Renfrew 1979). As a concept, it has also been used to explain sudden shifts within settlement patterns (Wagstaff 1978: 165–78; Renfrew and Poston 1979: 437–61; Wilson 1980; for a critical comment, see Baker 1979: 435–7).

Others have responded to the unsatisfactory nature of mechanistic or deterministic models of change in a different way. Curry, for instance, argued for the study of change in terms of chance and contingency. He called for a form of analysis that approached change from a wholly different direction. Instead of thinking that we can best understand change by defining the preconditions or determinants that lead to a specific outcome, we should start with the assumption that change is random rather than determinate and that any number of possible outcomes can be generated by it. In these circumstances, what the analysis of change should concentrate on is the task of defining the factors that constrain it, making one outcome – 'the most likely state' – more probable than others. Whereas with deterministic interpretations, he argued, we define a set of causes and generate a result that differs from actual conditions by an 'error term', with the formulation of change as a random process we start 'with unconstrained and independent random variables and, by introducing dependencies and constraints, achieve results of various likelihoods' (Curry 1972: 611).

More recently, such an approach has become associated with maximum entropy modelling. To date, such models have been restricted to specific rather than general instances of change and to predicting future rather than analysing past change. Yet as an approach, it has wide potential. This was certainly apparent to Curry when he argued that a powerful source of constraint was the accumulated investment in the physical structures underpinning social systems including settlement, transportation technologies and industrial plant (Curry 1972: 616; Curry 1964: 138). Jones's analysis of global economic growth in the very long term can also be seen as an attempt to model the role of constraint. All societies at all times, he argued, have a propensity for what he calls low-intensity economic growth. Whether such growth actually intensifies depends on whether other factors, such as governments, rent-seeking, etc., work to 'suppress it' (Jones 1988: 40). To this, of course, we could add the rigidities generated by the organizational and symbolic structures of a society. Invoking such constraints provides an analytical means by which the rigidities and inertia inherent in social structures and systems can be used to model societal change, enabling the emphasis to be shifted from what causes change to what constrains it.

Immanence models: life-cycle concepts
Taxonomically, we can distinguish another discrete group of interpretations that construe change as immanent in the nature of society. As a group, they can be further organized into two sub-types. The first covers those based on some form of life-cycle model. These see change as the playing out of a fixed potential, the unfolding of a tendency for society and its basic institutions to develop in a particular way or along a particular life-path. Needless to say, they draw heavily on biological analogues with stages of youth, maturity and old age. Such formulations are not common amongst social theorists but they have been widely used in economic models of change and in models produced by human geographers. What stands out from them is the simple fact that society or its institutional forms are seen as moving along fixed pathways of change. All the information or conditions needed to guide their development are present at the outset. In a sense, life-cycle models are a case of the more things change, the more they remain the same. At the end of a cycle, one is not presented with a different world. If the same conditions are replicated, so will be the cycle of development. It is an approach that reduces time as well as

space to a common experience: this is well demonstrated by Rostow's stages of economic growth, with their lead-sector concept based on eighteenth-century British experience being applied to less developed countries in more recent times. Indeed, in their extreme form, such approaches have led to ahistorical concepts of change. The epi-genetic models of societal change developed by Ekholm and by Friedman and Rowlands provide an important variant of the life-cycle model. They depict societal or socio-cultural systems as based not around isolated or single societies, entire unto themselves, but around systems of societies, or 'world systems', that are organized on a core–periphery basis. Within these 'world systems', each society comprises structures and processes that are developed around a mixture of both local and extra-local space. Over time, the size and complexity of these 'world systems' can change, whilst their geography is continually remapped with new systems erupting out of the periphery of old ones. However, their organization as core–periphery systems remains the same, hence their epi-genetic character (Ekholm 1980: 155–66; Friedman and Rowlands 1977: 201–76).

Immanence models: change as dialectics
Easily the most widely debated form of change based on immanence is that employing some form of dialectical change. Society is portrayed as containing inner conflicts that are present or immanent in its very nature as a society and which are resolved through revolutionary change (Claval 1989: 260–8). In its now standard form, such an approach has been widely used by Marxist analyses. Change is seen as triggered and driven forward by class conflict, with an array of other changes being seen as contingent on the resolution of this class conflict into new forms of class relations. The Brenner debate on the switch from feudalism to capitalism offers a good illustration of how Marxist analyses use such conflict to drive fundamental change. Rejecting views that anchor the shift from feudalism to capitalism in the disequilibrium that flowed from the growth of market activity and the *consequent* rise of a mercantile class, Brenner stressed the importance of a *prior* change in class relations. It was this shift in class relations and, therefore, in relations of production that created the context for other changes such as the growth of capitalism and the expansion of market activity. In effect, Brenner is internalizing change within social process (Brenner 1977: 61). Ekholm's version of 'world-system' theory also makes use of class conflict, using it as the means whereby new core systems emerge out of old peripheries at the expense of old core elites. Her reading emplaces the problem of class conflict geographically, treating it as a conflict between core and periphery. By comparison, Wallerstein's 'world-system' theory is inclined to see class conflict as following from the growth of trade and the unequal allocation of its benefits, both socially and geographically (Wallerstein 1974).

Change as social action

Interpretations which stress the primacy of human agency in the process of societal change form a diverse group. For some, change is about the action of human agents and human agents alone. In the hands of E. P. Thompson, such an approach becomes a rich interweave of human perceptions, intentions, actions and events (Thompson 1968). More so than any other type of approach, it represents an analysis of change in experiential terms, one that builds its understanding through the careful reconstruction and deconstruction of events and the role played by particular groups and communities.

Others accept that human agents and their actions are paramount, but feel that they cannot be divorced from the wider problems of structure. This concern to maintain the link between the actions of individuals and the wider structures of society has produced a large and complex family of models dealing with change. Amongst historians, it provided the basis for the core philosophy of the *Annales* school of history (Baker 1984: 1–27; Lloyd 1993: 117–27). Scholars like Braudel developed a history based on different scales of analysis. On the one hand, there were the short, even explosive moments of change, played out at the 'tempo of individuals' (Braudel 1972: 14) and studied through *l'histoire événementielle*, or the history of the event. Around these events, there is a larger scale of analysis based on the trends and cycles of social and economic life or what are called 'conjunctures'. At a still larger scale, that of the *longue durée*, are 'structures'. As Braudel put it, these are large-scale formations that have an integrated 'architectural' form (ibid.: 17) which, when seen from the vantage point of history, helps make us aware that because such structures 'encumber history, they impede and thus control its flow' (ibid.: 18). In effect, the link between the short term and the long term, between event and structure, is portrayed as if the flow of social change was like a lava flow, with those parts which had cooled and solidified obstructing and deflecting the surge of fresh eruptions. A similar balance is evident in the recent call by Lloyd for a structural history, but one which, whilst recognising the separateness of what he calls 'agential people' and 'institutional structures' (Lloyd 1993: 6), plays on their interaction, each sustaining the other (ibid.: 53–5). Yet despite their 'dialectical' interaction, he also acknowledges, like the *Annales* school, that there are circumstances in which institutional structures 'constrain' as well as 'enable' people, adding that institutional structures are 'relatively autonomous' (ibid.: 6) and have a 'powerful tendency towards maintaining temporal and spatial continuity' (ibid.: 64).

Amongst modern social theorists, a number have grappled explicitly with the question of how to relate human agency and human action to the wider problem of social structure. For Giddens, we should not try to reify structure in a physicalist way as if it were something with rules of its own. The structures of society are embodied in, and sustained by, the action of knowledgeable beings and not borne along carapace-like by 'mechanical dopes'. Through his concept of structuration, structure and practice are seen as combined and recombined *recursively*, with each being both means and outcome of the other. But being based on the actions of reflexive agents rather than mechanical dopes, both have movement or drift, as agents interpret and reinterpret the relationship between social practice and structure. This, in itself, instates change as a facet of the everyday routines of social practice. However, there is a larger point to be made. Once structure is penetrated by social practice, it ceases to be an exogenous or independent factor in societal change, one possessed of its own needs and constraints (Gregory 1981: 10). Instead, change becomes a product of the events and processes played out in specific circumstances. Social practices cannot be abstracted from their specific spatial and temporal context, their time–space distanciation. Though the shift from one type of society to another involves structural adjustments, it is not driven by structural needs nor does it have a 'mechanical inevitability' (Giddens 1981: 82). The actual process of change is an episode, usually occurring at time–space edges, and depending 'on circumstances or events that may differ in nature according to variations in context' (Giddens 1984: 245). Giddens's work has greatly influenced recent geographical thinking on change. Yet arguably, his own concept of change, at least that part of it which is based on 'episodes', is not linked to his concept of structuration in a wholly transparent or explicit way. Indeed, generally, structuration theory copes better with routines, with stasis, than with issues of change, despite its claims to the contrary (cf. Gregson 1986: 196–7).

For Giddens, society never cedes control over change to system needs. Faced with the same duality, Habermas sets a different balance of argument. He accepts that the objective analysis of society in systemic terms forms a level at which the problem-solving capacity and steering performance of a society can be evaluated. However, as noted earlier, the hallmark of his approach is the importance which he attaches to the structures of consciousness that constitute the lifeworld. For this reason, the starting point for his analysis of change lies with how societies rationalize their lifeworld, including their world view, as well as with the communicative competence of individuals to reach an understanding one with another. The most fundamental forms of societal change originate, in the first instance, from the qualitative changes that transform these 'structures of linguistically-produced intersubjectivity' and the norms and world views embedded within them. Though he concedes a 'development logic' to the way structures of consciousness change, a logic that embraces the increasing capacity of society to learn, rationalize and communicate, its phasing depends on chance historical events as much as on the working out of any inner potential (Habermas 1979: 98–9; Habermas 1987: 137 and 184). In his world of modernity, the world incepted by the Enlightenment, communicative rationality revolves increasingly around a decentred consciousness which differentiates between the natural, social and subjective worlds. These are part of the 'burdens placed on the internal structures of the lifeworld by growing system complexity' (Habermas 1987: 284; White 1988: 104–5). As modernity progresses, individuals are less dependent on customary or normative interpretations and are more able to use their individual competence for communicative action to produce a rationalized lifeworld (ibid.: 98). This affects the co-ordinating role of the lifeworld and its symbolic structures. It has given groups, especially marginal groups, a potential for the rationalization of their lifeworld in terms wholly different from that offered by dominant groups within the system as a whole. The conflict engendered by this legitimation crisis provides Habermas with a basic process of change at least for modern societies.

The work of the social theorist, Unger, provides us with yet another attempt at striking a balance between structure and agency in our understanding of societal change. It has similarities to Giddens's notion of structuration but is couched in different terms. His approach is founded on a firm rejection of so-called deep-logic approaches like Marxism. By positing only a limited number of paths and outcomes for social change, such approaches erode society's structural diversity, foreclosing its 'sense of historical possibility' (Unger 1987a: 93). In their place, Unger offers a reading of change that sees societal systems as pivoted around contexts whose potential for movement or change is caught between what he terms the 'push of sequential effects' and the 'pull of plasticity' (Unger 1987b: 277; Unger 1987c: 212). What he terms a 'formative institutional and imaginative context' helps structure everyday routines, biasing their outcomes towards particular 'sequential effects' (Unger 1987b: 58). As in Giddens's work, though, the very devices that stabilize such formative contexts can 'endlessly produce the occasions and instruments of destabilization' (ibid.: 277). For Unger, these 'occasions and instruments' depend on what he calls 'less entrenched contexts' which enable society to question and experiment with the institutional and organizational forms of society. These less entrenched contexts, he proposes, are those that undermine 'rigid roles and hierarchies' and bring 'framework-transforming conflict and framework-preserving contexts closer together' (ibid.: 249), enabling formative contexts to 'be challenged in the midst of ordinary life' (ibid.: 280). At the very heart of his argument is the proposition that what matters for change is not the transformation of whole systems, or the 'false necessity' of preserving existing ones, but the extent to which societies can experiment with new routines and institutional forms and the extent to which these new forms and institutional forms can be recombined in a continuous exercise of disentrenchment (Unger 1987c: 208; see also, Nelson and Winter 1982: 130). Society's capacity to maintain an open agenda on its institutions and their combinations represents for Unger the pull of negative capability (Unger 1987b: 277 *et seq.*).

Just as approaches based on human agency and human action represent themselves as counter-arguments to those which stress the mechanistic or deterministic influences of structure, so have systemic approaches also responded to this criticism by building models which attempt to strip away their deterministic and inertial qualities and incorporate an ongoing flexibility and responsiveness to social needs and intentions. In simple terms, this was done by incorporating more sophisticated sensors to monitor threats and opportunities and by defining regulatory systems that are capable of positive as well as negative feedback, the latter being used to conserve what is deemed desirable and the former being used to institute continuous qualitative change in response to new opportunities and possibilities. Societies are invested with order but not at the expense of their capacity to make ongoing, even continuous change. This hybridization of static and dynamic perspectives was used by Talcot Parsons. Society, he argued, achieves stability through the progressive differentiation and integration of roles, a 'constant adaptive upgrading' (Parsons 1966: 22), a process designed to achieve greater adaptation between society and its opportunity space. As the system works towards greater integration and complexity, its value pattern must incorporate 'higher and higher levels of generality in order to legitimize the wider variety of goals and functions of its sub-units' (ibid.: 23). As Smith observed, trying to explain this 'change-within-order', Parsons creates an argument that is hardly different from one that allows for 'order arising out of change and change itself proceeding in an orderly fashion' (Smith 1973: 131). Change was seen as no less continuous by Sorokin, 'an inherent property' of all societal systems (Sorokin 1967: 68). It was something immanent, arising out of the properties of, or conditions within, the system itself, so much so that we can regard change as realising 'its inherent potentialities' (ibid.: 69). Yet far from being determinate and predictable, its continuity and smoothness was varied by changing rhythms and tempi. More significantly, each phase of change laid down new conditions for subsequent change so that we are faced with a process that is cumulative, 'a system incessantly transforming both itself and its milieu' (ibid.: 69).

Attempts to see this continuous capacity for change in a three-dimensional way and to do so systemically are well illustrated by Isard's elaborate attempt to model what he calls time–space transitions. His model concedes some equilibrium to society, though not the metastable phases of equilibrium envisaged by Butzer. The differences between them lie in the way they each negotiate round the fundamental opposition between old and new. Whereas Butzer saw early Egyptian society as making fresh adaptations in response to catastrophic conjunctions, Isard allows modern society a more consciously creative part in the process. Each 'new regime', he suggested, 'is by necessity one that makes use of the old regime while changing it' (Isard 1977: 6). Each new state is 'a state which is accessible without major threat of destruction, from the old regime' (ibid.: 9). Isard claims the impossible, a society 'organized not simply to maintain an efficient steady state but rather to undergo an efficient process of change over time' (ibid.: 6). He is arguing for how society should ideally be organized rather than how exactly it is. On the one hand, he has society making an adaptation. In doing so, it selects against variation and deviation and, in the process, achieves what he calls a fitness peak. On the other, however, he argues that in order to survive, it must maintain some variety. When these deviant sub-systems produce answers or adaptations that offer a better fit to new environmental conditions, then such variation is selected as the system passes its fitness peak in respect of earlier adaptations.

Isard tries to rationalize his use of two potentially conflicting models by describing society as a complex adaptive system. Yet his use of this concept differs materially from its use by social theorists like Buckley. In the hands of Buckley, the idea of society as a complex adaptive system is used to produce a reading of society as a continuously responsive system. Whereas Butzer and Isard, despite their other differences, frame the problem in terms of phases of standstill and stability punctuated by phases of change and adjustment, Buckley offers a process of change that is more continuous and incremental. Society is set up as a system which is open externally and internally to flows of information. Feedback control loops, he argued, make possible both self-regulation and self-direction, the latter enabling the system to 'change or elaborate its structure as a condition of survival or viability' (Buckley 1968b: 490). Instead of seeing society as composed of large structural units like institutions or cultures, Buckley's society becomes an ongoing 'morphogenetic process whose form or shape is determined by a systemic matrix of interacting, goal seeking, deciding individuals and subgroups' (ibid.: 497). His intention here is clear. He is shifting the emphasis from what he sees as static to what is potentially far more responsive and dynamic. At root of his approach is a basic tenet of complex system analysis, that is, that a system state is not about the determinate playing out of some initial conditions but, as with Sorokin's interpretation, is something realized through the experience or history of the system. Whereas closed systems tend to collapse when faced with new information from outside, open systems are able to reorganize. Thus, as a complex adaptive system, society is able to transform or elaborate itself as a condition of survival. For Buckley, the term 'morphogenesis' captures this self-transforming process (Buckley 1967: 58–9). Change is not something exceptional or intrusive but part of the way society is organized, a process whose potential changes through change.

Amongst recent literature, the most explicit use of the idea that society is a complex adaptive system has come from Aulin. Whereas simple cybernetic or homeostatic systems seek to be self-repairing, always trying to correct deviation back to a fixed goal, complex systems are self-steering. They have the capacity to respond to a succession of differing opportunities or demands.

This capacity for steering themselves in response to random as well as regular flux in their environment means they have the capacity not simply to change but to go on changing. In effect, they have the means to change continually without ever recovering past states. For Aulin and others, this answers the needs of systemic and historical approaches to the question of change. In order for a complex adaptive system to survive and to cope with change, it has to match the variety that exists in its environment with an equal variety of control (Ashby 1968b: 129–36). Of course, in actuality, societal systems have imperfect knowledge and are invariably organized in ways that are far from optimal when it comes to coping with environmental flux and change. For this reason, there is always an uncertainty over regulation so that the disorder threatened by environmental change, with its increase in entropy, is always greater than expected. Aulin sees society as coping with this dilemma in two ways. First, it can reduce the uncertainty of regulation by developing a greater hierarchy of control, thereby improving the system's controllers. Second, uncertainty can also be reduced by simply improving the sensing and monitoring of the environment, improving the system's regulators. For Aulin, these two types of change work against each other in that more efficient regulators reduce the need for more elaborate or hierarchical controllers. Precisely which sort of response is emphasized depends on prevailing conditions. Where or when survival is at issue, one is more likely to find an emphasis on increasing control hierarchy and vice versa. There is, however, another conclusion to be drawn. In so far as a preponderance of the former produces a self-steering system and a preponderance of the latter produces a self-organizing system, it follows that societal systems move between limits set by an excess of either. Put in everyday terms, he is defining how societies vary historically and geographically between systems that are strongly co-ordinated, hierarchical and conservative and systems that are loosely co-ordinated, have little hierarchy but have far greater scope for self-steering (Aulin 1986: 100–18).

Marginal or peripheral sources of change

A third category of change is made up of those interpretations that stress the role of peripheries and margins, or of those groups and communities that inhabit such spaces. Having already touched on some of these ideas in the opening chapter, only a brief review is needed here. For some, margins or peripheries, whether defined in relative or absolute terms, are spaces in which conjunctions occur. By virtue of the very fact that they are margins or peripheries, or 'in-between space' as Bhabha put it, they are invariably hybrid, a place where differences can be brought together and, through their mediation, 'newness' encountered (Bhabha 1994: 36). As well as experiencing change through the potential offered by their liminality, their role as the space where one society gives way to another and where wholly new values and symbols

Table 2.3 *The product of societal change*

1. Change by expansion or contraction (a–A, A–a)	2. Change by replacement or substitution (a–b–c)	3. Change by incorporation or aggregation – Upwards $(a+a+a=B)$ Outwards $(a+a+a=A)$	4. Additive or accretionary change (a–ab–abc)	5. Involuntionary change $(a-a^1, a^2, a^3)$

can be forged from their opposition and interaction to dominant core areas, margins or peripheries can foster change through their role 'as a site of radical possibility, a space of resistance' (hooks 1991:149). This is a theme well articulated by Mann in his analysis of socio-cultural change in the Middle East over later prehistory and the early historic period. Through successive changes, the margins were instrumental in driving change, showing an endemic capacity for what Mann calls 'interstitial surprise' (Mann 1986: 539). A similar theme provides the process which energized Ekholm's epi-genetic model of socio-cultural change, with new core–peripheries continually emerging out of former peripheries (Ekholm 1980: 155–66). Arguably, it also provides some of the dynamic for Giddens's notion of change taking place around time–space edges.

Defining the product of societal change

At first sight, questions about what societal change amounts to, its sum product as a process, would appear to raise few conceptual difficulties. In fact, many studies of what change amounts to have been shaped around fundamental ideological presuppositions and, as a consequence, are restricted to particular outcomes. The Marxist assumption that change is a revolutionary process, always preceded by revolutionary conditions, is an obvious case in point. Yet viewed taxonomically, meaningful change can take a variety of forms (see Table 2.3). Fundamentally, it can amount to quantitative or scalar change, with patterns or systems simply expanding by a process of diffusion or contracting by a process of shrinking, or it can involve qualitative or systemic change, with a change in the way patterns, structures and systems are constituted, organized or conceived.

If seen strictly, the former – quantitative change – can amount to a very simple even trivial form of change. The same patterns or systems either grow larger (a^1 being replaced by a^2, a^3 and so on) or they become smaller (a^3 being replaced by a^2, a^1, etc.). However, in most cases, such change is merely a phase in the unfolding of qualitative change, with the growth of particular patterns and systems replacing others or creating conditions, via scalar stress, out of which a reordering into a qualitatively different system might arise. When we confine it simply to those circumstances in which change involves no more than quantitative change, then it ranks as a very limited form of change.

As regards qualitative change, we can differentiate between change by replacement, incorporation or aggregation either upwards systemically or outwards geographically, by additive or accretionary change and, finally, by involutionary change. Despite this variety, a large section of the literature takes a narrow view of qualitative change. New systems are produced by the transformation of old ones, *a* becoming *b* and *b*, in time, becoming *c*, etc. Change is an act of replacement or substitution, as much a destructive as a creative act. Such change underpinned both Turner's frontier thesis (Turner 1921) and Whittlesey's theory of sequence occupance (Whittlesey 1929: 162–5; see also, Mikesell 1976: 148–69). Of course, the free use of terms like the Urban Revolution, the Neolithic Revolution, the Agricultural Revolution and the Industrial Revolution are ample testimony to the deeply held belief by many that if change matters, then it happens in a revolutionary moment or phase, with one system being overwhelmed by another *in situ*. Yet despite its easy exemplification, and despite its powerful ideological underpinnings, this is far from being the only or even the dominant type of qualitative change. In some cases, the type of change involved is best described as integrative, with the upward integration of larger areas and more units leading to more complex forms of organization. Both Renfrew (1975: 3–59) and Isard (1975: 113–24) have explored instances of this type of change, with a cluster of similar units (*aaaaa*) giving way to *aaaaa+b*, where *b* is a higher order centre of information control. In others cases, qualitative change can be seen as seeking out and exploiting interstitial areas, evolving new forms that accrete themselves to older-established forms or establish wholly self-driven forms. Such forms can be expressed as a compound form of change, with *a* giving way to *ab*, then *abc*, with new forms always being appended to existing systems, or they can be represented as a form of change whereby *a* becomes *a + b*, then *a + b + c* and so on with new forms being established as separate systems. This sideways or crab-like shift towards change was first articulated by Sahlins and Service, who embodied it in their principle of geographical displacement (Sahlins and Service 1960). Its clear geographical implications were pointed out by Newson (1976: 239–55), whilst more recently, historical case-studies of real-world examples have been provided by Mann (1986) and Dodgshon (1987: 240–86). As a type of qualitative change, involutionary forms are difficult to vouch for, yet arguably, they can be aligned alongside revolutionary change as a way of absorbing growth internally, with systems being elaborated in response to growth. Usually, change is deflected towards an involutionary, as opposed to evolutionary, solution owing to the circumscription or caging of growth as with Mann's reasoning out of how the first complex societies evolved in the valley-based ecologies of Mesopotamia (Mann 1986: 74–5).

One final hybridized form of change, hybridized because it is neither wholly qualitative nor quantitative in character, involves change by reduction, with a diverse population of systems being thinned so as to produce a single or small cluster of dominant systems, *abcde* becoming, *abcd*, *abc*, *ab* then *a*. Much of the debate about globalization can be seen as a debate about the erosion of diversity, with regional cultures and economies being overwhelmed by the predatory expansion of a 'successful' capitalist or global system. Locally, such a change could be read as an example of change by replacement or substitution (*a* becoming *b*, then *c*, etc.), but seen in its broader context, it clearly involves a different type of change. In the case of the former, the new emerges out of the old. In the case of the latter, the new is something that displaces the old.

The morphology of change

Viewed temporally, all change possesses a morphology or trend. Three types of trend are distinguishable (see Table 2.4). First, there are those instances in which change unfolds continuously or gradually. An illustration of this sort of change is provided by Buckley's notion of society as a continuously adaptive system (Buckley 1968b): such change appears as a smooth, linear curve. Usually though, such arguments cover only limited phases of societal change. Conceptually, they can be linked most closely to development or immanence models that assume that change is the realization of an inner potential. Each step or stage is linked to the previous stage, simply because the character of change is incremental or marginal. If sustained, though, such change will ultimately generate scalar stress and the need for more radical, qualitative change if stagnation and decline are to be avoided. By its very nature, such qualitative change is rarely a smooth or sustained trend.

Providing a second type of trend are those arguments – collectively, a less coherent group – which favour some form of broken trend, one that involves revolutionary moments, knickpoints, punctuated equilibrium, non-linear processes, bifurcation points, dissipative structures, catastrophic collapses or sudden upward/downward surges. Morphologically, these are all similar in that they treat change as a non-linear or discontinuous trend, with society or its major sectors jumping or collapsing between different states, whether we define those states in terms of normative values, organizational forms, production functions, or whatever. Some restrict outcomes at the point of change to a few possibilities, whilst admitting a wide range of outcomes. Some see outcomes as in some way determinate, based on what had gone before, whilst others, especially those based on non-equilibrium models, assume even minor fluctuations or marginal processes can be amplified unpredictably into new

Table 2.4 *The morphology of societal change*

A: Continuity–discontinuity debate

1	2	3	4	5
1. Change as a smooth, continuous trend	2. Change as a non-linear trend: including revolutionary moments, punctuated equilibrium catastrophic switches between system states, bifurcation points and dissipative structures	3. Continuity/discontinuity as a time–space trend – role of diffusion – types of diffusion: expansion, relocation, contagious, hierarchic (inc. cascade)		

B: Biological analogues

1	2	3	4	5
1. Evolutionary – progressive evolution, anagenesis, ontogeny	2. Evolutionary – macroscopic diversity through error copies, etc. + natural selection	3. Evolutionary – sideways or branching speciation	4. Involutionary – change by elaboration	5. Developmental – epi-genesis

equilibrium states or dissipative structures. But however constituted, what unites such interpretations is the basic assumption that the trend of change is broken or punctuated rather than continuous. Also worth noting is that in most cases, these non-linear trends are seen as irreversible. The main exceptions are those which use catastrophe theory. This is one of the reasons why Renfrew employed such concepts to model the transition between chiefdoms and early states for, initially, so-called proto-states were prone to collapsing back into chiefdoms (Renfrew 1979: 418–507).

A third type of trend captures the three-dimensional nature of societal change, highlighting the fact that it works itself out through space as well as time. Geographers, of course, have long dealt in time–space forms of change through work on diffusion. Classification of the different types of diffusion can be used as a guide to how such time–space trends can be further classified: expansion diffusion (including its sub-types of contagious and cascade or hierarchic diffusion) and relocation diffusion. However, more often than not, such models are intended to describe how different artefacts or traits of socio-cultural change spread rather than to account for how change originated. Yes, there is a distinction drawn between hearth areas and areas of expansion, but this is different from saying that diffusion automatically provides an intrinsically geographical model of why change originates where it does. For this reason, those time–space models of change which try to incorporate such an explanation can be regarded as defining a distinct morphological type of change. In effect, such a morphology actually combines the smooth trend of a developmental model with the discontinuity of a revolutionary model by separating them geographically.

Occupying a position mid-way between those which define the sum effect of change and those which define its morphology are attempts to classify change in terms of its biological analogues. On the one hand, such analogues have been used to describe the product or achievement of change, with new societal forms or institutions being likened to the emergence of new species by progressive evolution, modification by descent, mutation, natural selection or elaboration. On the other, such analogues have also been used to describe the form or trend of change, likening it to ontogenesis, anagenesis, punctuated equilibrium, sideways or branching speciation and epi-genesis. Though some have criticized evolutionary approaches to societal change as an inappropriate and flawed perspective (e.g. Gellner 1988: 2–3), the use of biological analogues is deeply entrenched as much in the work of social scientists (e.g. Nelson and Winter 1982; Laszlo 1982; Mokyr 1990) as in that by biological scientists talking about social evolution.

I have tried in this chapter to establish the range of debate that exists over societal or socio-cultural change. As one might expect of such a fundamental problem, the different viewpoints that exist over its interpretation are many and varied. Some of these differences stem from the simple fact that supposedly generic interpretations actually address confined sections of the problem, though this has not stopped some that deal with different aspects from being opposed to each other as if they were competing interpretations of the same aspect. Of the real differences that remain, the prime disagreement is between those which see society as typically organized for stability and those that see it as organized for change. Yet even within the existing debate, there are ideas that provide us with a way not so much of removing these differences but of framing them in such a way as to make them consistent parts of the same argument. The ideas which I have in mind are those which treat the problem of change not simply as a temporal process but as a time–space process, that is, as having spatiality. Gregory has recently called for a 'contextual theory of society' (Gregory 1990a: 217–33). This could equally well be transposed into a call for a contextual theory of societal change, one that is sensitive to the different ways in which society's capacity for change varies geographically across society and between societies.

3

The experience of change: world systems and empires

> It is a matter of probability, not of organismic inevitability, that cultural systems, like all human institutions, will eventually collapse given a sufficiently long span of time.
>
> (Butzer 1980: 522)

It was noted in the previous chapter that the balance of recent debate over societal change has shifted towards the view that society is, by nature, open to change. For some, societies are capable of negotiating change as a sustained ongoing process of self-transformation, of being 'continuously responsive'. Even amongst those who reject the idea of society as a directed or self-steering system, some would accept that change is at least 'reflexively monitored' and, therefore, needs to be examined through the role of human agency and not bound deterministically within structures, organizational constraints, system needs and so on. Whatever their differences, a common denominator between these otherwise different types of approach is that society is capable of change. Whether change is anticipated, a consciously planned process of adjustment, or simply an opportunistic response to crises or unforeseen circumstances, society is seen as having both the means and the will to cope.

In reply, I want to argue that if we set the theoretical debate beside the historical record of societal change, both long term and recent, it provides only qualified support for such assumptions. On the record of the past, society cannot be regarded as organized for easy change. Nor can all aspects of change be seen as encompassed within an interpretation that stresses the 'reflexively monitored' nature of change. Instead, I want to argue that the lesson to be learnt from the historical analysis of societal or socio-cultural change is that whilst it is an endemic feature when seen at an aggregate or gross level, nevertheless, the experience of social systems when considered individually is that they *ultimately resist ongoing change or cope with it badly*. In this chapter, I want to review the problem at the scale of large-scale systems, a scale represented by empires and world systems. Particular attention will be given to the stresses and strains experienced by such systems and how these affected their long-term capacity for change.

Empires and world systems in time and space

By definition, an empire is an extended and hegemonic projection of power by a core or dominant society so as to embrace societies that are extra-local to it, or outside of its own immediate territory. Some of the classical empires, notably the agrarian bureaucracies of the Near and Far East, integrated societies that were broadly uniform in their social, economic and political character. However, some commentators see empires as being, in essence, anything but homogeneous (Gilpin 1981: 110–11; Watson 1992: 125). The main problem lay in the realities of over-reach, with empires always reaching out to secure their frontiers, and drawing in or interacting with societies that were ultimately different in character, so that most empires had a periphery or marcher area that was made up of societies – usually subsisting under different ecological conditions (Lattimore 1940: 350–5) – that were perceived and treated as different, a periphery that fostered a sense of the other. As with other forms of core–periphery system, empires usually involve an asymmetrical structuring of power between a metropole or centre and a periphery, with the nature of authority thinning and metamorphosing with distance from the core (ibid.: 352–3; Watson 1992: 126; Doyle 1986; Hall 1988: 20–1). Of course, where attempts were made actually to draw a line, such as with the various walls that fronted different Chinese empires to the north and north-west or Rome's north-western frontier, the outer edge of an empire could involve a very intense and focused concentration of power.

The formation of empires has been explained in three ways. For some, the explanation is to be sought in the nature of metropoles or emergent centres of power, and the eruptive way they acquire a disposition to expand their authority or hegemony over an ever increasing domain by both formal (i.e. political alliances) and informal means (i.e. military conquest, trade). For others, the answer lies in the marcher processes or circumstances that tend to exist around the edge of strong systems, with cross-border conflicts or threats being resolved by a process of agreed or forced incorporation and subjugation. Finally, there are those who root the formation of imperial structures in the rationalist nature of socio-political systems, with strong systems tending to expand as a mark of their superiority relative to other rival systems (Doyle 1986). Though some have contrasted the extent to which early empires were assembled using military force with the way more modern forms have relied on the integrating power of trade and exploitation, any exclusive distinction between these two forms of integration is difficult to maintain. Early empires put together by brute force invariably carried heavy maintenance and administrative costs that were paid for through the levying of taxes and tribute and by terms of trade that were favourable to the core and its dominant groups. Their primary and enduring dependence on the output of agriculture, and the difficulties of raising output except through costly schemes of land improvement like irrigation, had this consequence. As the maintenance costs rose, so too did the pressure for an extension of territorial control or their resource base (Gilpin 1981: 110–12). In other words, their integration was as much economically driven as it was politically or militarily based, with cores being defined as much by the accumulation of capital as by the concentration of coercive power (Ekholm and Friedman 1993: 59–80). In essence, what distinguished empires was the way in which they created physically extended structures that embraced a range of territories and societies around a dominant core or group: this structuring of power became the basis for a system of exploitation and unequal exchange between the core and its dependent societies. Their expansive projection of power and exploitation invariably created problems over how huge areas of space and resource could be integrated. For the more successful, the solution usually involved innovations and developments in military techniques, bureaucratic skills and administrative concepts, and in the storage and communication of information, the latter ranging from new media of storage and communication (i.e. writing) to the physical means by which information was moved from one part of the imperial realm to another (i.e. the wheel, road building and so on). In fact, though territorially structured, empires were, as one writer put it, 'radially managed', meaning that the exercise of power via available authoritative and allocative resources was structured along well-defined routeways, linking one centre of control to another (Watson 1992: 125–6). Indeed, for writers like Mann, the arterial rather than areal deployment of these imperial resources of control in the form of overlapping networks of interaction provides the only means by which they and other polities can be effectively defined in a social as well as geographical sense (Mann 1986: 16). But as Innis might have observed in reply, the reality of 'large-scale political organization implies a solution of problems of space in terms of administrative efficiency and the problems of time in terms of continuity' (Innis 1972: 170). Put simply, their integrity as imperial systems depended as much on their ability to rule well as on their ability to conquer (Lattimore 1962: 105; Hall 1988: 20–1). The former rewarded the ability to spread power and the latter, the ability to concentrate it (Mann 1986: 142).

Not surprisingly, because of their limited solutions to the challenge of ruling and surviving in space and time, all empires ultimately proved vulnerable rather than continually responsive to the threat of change. For this reason, when seen overall, their collective histories are characterized by flux and discontinuity. This imperial cycle, as it has been termed, is rooted in the structural problem created by the sheer scale of their growth and its changing relationship to administrative and military costs. The latter follow a 'U-shaped' curve, first declining through economies of scale and then increasing. By comparison, the intensity or closeness of control follows an 'S-shaped' curve, with an initial increase in the intensity of political control being followed by a decrease as expanding reach or distance, rising administrative and military costs, and the logistical difficulties of sustaining expansion conspire to weaken imperial control (Gilpin 1981: 107 and 115). Ultimately, the decline phase of the imperial cycle follows from the failure of empires to control this 'U–S' divergence as costs rise but effectiveness of control diminishes. The potential contradiction has been put in a more neo-Malthusian way, with a tendency for administrative and military costs to rise geometrically whilst the resource base can, at best, only respond with arithmetical growth (Elvin 1973: 110; Gilpin 1981: 148).

In recent years, the debate over empires has broadened and now overlaps with the debate over world systems. As a concept, world-system theory was initially developed around the capitalist world system which Wallerstein saw as first emerging during the 'long sixteenth century' (Wallerstein 1974). The capitalist world system is a system of order based on unequal exchange, with core areas exchanging high-value manufactured goods for low-value primary products, raw materials and food from peripheral areas. Though typified in terms of trade and market exchange, Wallerstein saw the capitalist world system as determining the wider character of countries, with core areas and their social hierarchies developed around democracies, free labour systems, high levels of investment and low rates of return, whilst peripheral areas were characterized by a lack of democracy, coercive labour systems, low levels of investment and high rates of return (see Fig. 3.1). Though he acknowledged a state of semi-periphery in between, this was Wallerstein's way of allowing for the possibility that countries could move between the core and periphery. But always, some countries would form the core of the system and others, the periphery.

Recent debate has greatly extended the discussion over world systems by arguing the case for the existence of world systems long before 1500. Moreover, some have argued that these pre-1500 systems were no less capitalist in their constitution than Wallerstein's post-1500 world system, with core areas accumulating capital through their exploitation of peripheral areas via unequal exchange, tribute exaction, etc. (Ekholm and Friedman 1993: 59–80). Indeed, these pre-1500 world systems are seen as extending back to the first civilizations or empires of the third millennium BC (ibid.: 59–80; Frank and Gills 1993: 3–55). The link, if any, between these early world systems and Wallerstein's post-1500 world system is still contested. For some, each world system is a separate system, one that evolved under different conditions. Abu-Lughod, for instance, has stressed how a world system centred on the Ottoman to north India region over the late thirteenth and early fourteenth centuries had decayed before that based on Europe had begun to emerge (Abu-Lughod

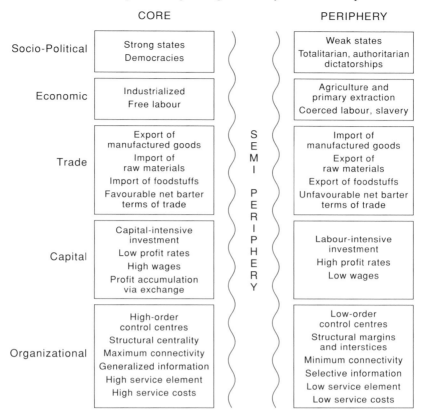

Fig. 3.1 How core and peripheral areas differ within a capitalist world system. Based on Dodgshon 1993: 28.

1989; 1993: 288–9). She accepts that sub-regions could persist from one system to another, but overall, each world system is a restructuring of socio-economic order. Others draw attention to the fact that whilst each world system is a restructuring, many appear to arise from amongst the regions and societies that existed in the periphery of prior world systems. Still others couch the problem in the singular. Building on evidence showing the extent of long-distance trade from the third millennium BC onwards, with the stability of core areas and their hierarchies in widely scattered regions fluxing in step with the growth and decay of such global trade patterns, they argue that there has only been one world system then and since. As a development of this latter view, Gills and Frank have calculated that this singular world system experienced eight separate cycles of growth and decay, each phase of growth or decay lasting between 100 and 200 years (Gills and Frank 1993b: 188).

One of the reasons why such disagreement has arisen over this problem of whether there has been just one long-lived world system or a succession of different systems arises partly from the fact that from the outset, world systems were not structured in a simple way. What we label as core could consist of a hierarchy of major and lesser centres, or hegemons, diffusing the status and function of the core out over a wide area and enabling it to operate at different scales. Under these circumstances, it would be easy for successive configurations of a single world system or successive world systems to overlap to a point at which their separate identity might be confused, or at least be open to debate.

The early empires

The Middle Eastern civilizations

From the moment the first civilizations emerged as close-knit networks of centralized, bureaucratically organized and hierarchical societies, based around locally dominant cities and urbanized religious centres, their constituent parts displayed a tendency to fall out with each other or to shift political and geographical allegiances under the slightest pressure (see Fig. 3.2). This was a feature of the very first civilizations that emerged in Mesopotamia. The earliest of the Mesopotamian civilizations was the Sumerian which developed between 4000–3500 BC in the lower valleys of the Euphrates and Tigris, an area in which alluvial soils provided high levels of fertility but where high levels of water control were needed to offset the semi-arid climate of the area (Crawford 1991: 18–19). What distinguished Sumerian society as a civilization was the greater complexity of organization, a complexity manifest in its ordering around city-states, each controlling a territory of agricultural land but each dominated by a single city and its needs. In character, Sumerian city-states like those around Eridu, Uruk and Ur were temple-based, with both their society and economy dominated by the demands of a temple complex dedicated to the city's god and serviced by a large community of priests and clerks. Each city's temple economy drew large amounts of food as tribute or render from its surrounding hinterland (Maisels 1990: 154). Indeed, so dominant were these needs, so critical to the state's organization, that some agricultural land was actually communally owned. Significantly, the development of early forms of cuneiform writing in Sumerian city-states also reflected the needs of the temple economy, with tablets recording the flows of grain and other agricultural products through the temple's system of storage and consumption.

The Sumerian city-states interacted through economic rivalry and straightforward warring, with pre-eminence shifting between them, but they were never welded into a single system or hegemony so that we cannot strictly

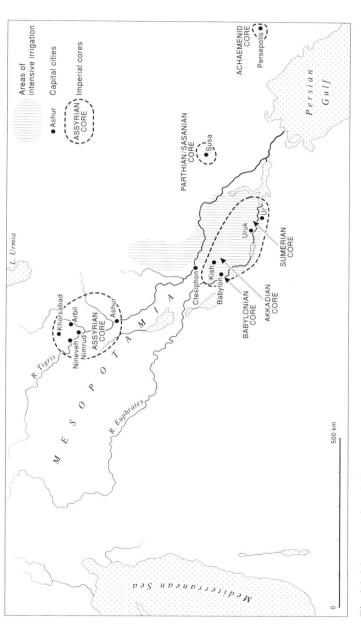

Fig. 3.2 Mesopotamian civilizations and empires. This map shows the shifting foci of Mesopotamian civilizations and imperial structures, identifying the core areas of the main civilizations or empires together with their capital cities. The early civilizations, such as those based around Sumerian, Akkadian or Babylonian rulers and cities, occupied the southern Mesopotamian or lower Tigris and Euphrates area. The later Assyrian Empire occupied both the lower and middle Tigris and Euphrates area plus most of the area depicted on the map that stretches westwards to the Mediterranean. The Parthian and Sasanian Empires based around Susa occupied the lower Tigris and Euphrates area plus the Zagros mountains and Iranian Plateau stretching eastwards across the right side of the map. Finally, the Achaemenid Empire, whose main administrative capital lay at Persepolis, covered the entire area shown by the map, and much more to the east. Based on data drawn primarily from Whitehouse and Whitehouse 1975: 68–71; Barraclough 1984: 54–7; *Past Worlds* 1988: 124–5, 156–9 and 186–7.

see them as forming an empire (see Fig. 3.2). In time, though, there appears to have been a shift away from the temple as a focus of political power to royal dynasties and their palaces. Over the closing centuries of the third millennium BC (2371 BC+), a dynasty based at Agade, close to Babylon, managed to weld Mesopotamian city-states into a more unified system which arguably formed the world's first empire: the Akkadian (2371–2230 BC). The Akkadian was short lived, with power passing back to the various city-states to be replaced by an even stronger hegemony under a royal dynasty at Ur (*c.* 2112–2004 BC). Yet this too was to be ephemeral, collapsing by the start of the second millennium BC. With the arrival of fresh Semitic-speaking groups in the form of the Amorites from eastern Syria, the city-state system was once again both reinvigorated with new dynasties and remapped with new centres of dominance. After a further brief eruption of unity under the hegemony of Babylon, the southern Mesopotamian region finally lost its momentum, the focus of power shifting northwards and eastwards. With its surges of more centralized, integrated power structures followed by phases of disaggregation, and with its repeated shifts in political focus, the region highlights the geographical flux of socio-political and socio-cultural order that was to be typical of subsequent civilizations and empires. However, such changes were only part of the problem.

A feature of the Mesopotamian civilizations was their close dependence on some form of water control, both to control the natural and often catastrophic flooding of rivers like the Euphrates and to extend their waters in a more controlled and productive way by means of irrigation. This is why Wittfogel labelled civilizations like the Sumerian as hydraulic civilizations (Wittfogel 1957a). Having in mind the fact that all the major early civilizations (i.e. Sumerian, Babylonian, Egyptian, Indus and Chinese) relied on elaborate systems of water control and irrigation, he saw the control, co-operation and administrative expertise demanded by such systems as the prime bonding factor for the more complex societies – hierarchical in structure and bureaucratic in character – that emerged with the first civilizations (Wittfogel 1957b: 343–64). In fact, the debate over the precise role of irrigation is far from settled. Reviewing the results of further archaeological work, Adams concluded that 'there is nothing to suggest that the rise of dynastic authority in southern Mesopotamia was linked to the administrative requirements of a major canal system' (Adams 1966: 67–8). Admittedly, there are clear signs of a dramatic population 'implosion' as settlement in the countryside was abandoned for larger centres like Ur and Uruk (Gibson 1973: 449), but we cannot link this surge in city growth and its implications for societal complexity with the *de novo* development of large-scale, co-ordinated water control systems. In fact, where the archaeological record offers a deep perspective on the problem, such as in the deeper alluvial soils of central and northern Mesopotamia, it suggests that initially, irrigation was developed locally through small-scale schemes organized at a village or inter-village level, schemes that were fitted in closely with the natural hydraulic regime of rivers (Adams 1966: 68). Though some would still disagree (Oates 1977: 481), the balance of opinion is that the realization of such small-scale schemes did not necessarily require, or induce, a shift to societal complexity as a pre- or co-requisite. Indeed, more recent interpretations have shifted the emphasis away from irrigation as the prime mover, drawing in other variables to what is now a more complex equation (Maisels 1990: 191–2).

Yet despite this broadening of debate, the ecology of the region still has a place in explanations. Thus, a theme running through a number of recent views is the extent to which the semi-arid conditions of the region confined agriculture to the land on either side of river channels, thereby circumscribing population growth. Whether we see this population growth as self-generated or as a factor whose significance was accentuated by inward migration from marginal areas, some see its natural circumscription or 'caging' as a vital input to any model of how societal complexity evolved in the region (Gibson 1973: 448; Mann 1986: 75). The circumscription of growth had other effects. One of the prime agro-ecological features of the region was not just the extent to which farming depended on river-fed irrigation schemes, but also, the basic instability of such a system. Radical shifts in the layout of river channels, and the progressive salinization of soils caused by the regular flooding of soils in areas where drainage was poor but evaporation high, fundamentally altered the geography of production between different areas (Crawford 1991: 37) and may have been a powerful contributory factor to the changing geography of power evident in Mesopotamia over the fourth and third millennia BC. It may even be the case that whilst the initial adoption of water control schemes in the region did not raise any special demands on existing forms of social organization, the need to devise co-ordinated and effective coping strategies when faced with the destabilization of these established schemes may have posed problems of a wholly different order.

There are comparisons to be drawn here with early Egyptian civilizations. The civilizations that arose in the Nile valley *c*. 3000 BC and which survived through various dynasties down to the mid-second millennium BC developed around a similar bounty of fresh alluvium and plant nutrients provided by the annual flooding of the Nile. As with the civilizations of the Tigris and Euphrates, it is tempting to see this basic ecology as providing a fundamental source of continuity across many socio-political changes. Indeed, the annual flooding of land and the renewal of the soil's fertility was taken as the paradigm for the origins and maintenance of life itself, with Egyptian cosmology seeing order, cultural no less than physical, annually renewing itself first as islands then as whole landscapes out of the waters of chaos as the latter receded (Wilson 1949: 39–70). Yet for all the enduring realities of the Nile as an essential source of fertility and belief, it was also a source of instability.

Butzer has drawn attention to the way in which phases of major political change were associated with periods during which the flooding of the Nile was unreliable. His argument is not that this unreliability precipitated change in itself but that it coincided with other changes, their concatenation being the catalyst of change. Thus, the couple of centuries leading up to the Old Kingdom (2760–2225 BC), the first Egyptian civilization, were marked by low water levels that inhibited flooding over large areas. Likewise, the end of the Old Kingdom itself was associated with failures of the Nile to flood widely, failures that coincided with the breakdown of central control and economic decline. The emergence of the Middle Kingdom (2035–1668 BC) brought the creation of a stronger, more centralized bureaucracy, the draining of the Faiyum Depression and greater government responsiveness to famine, but serious excess flooding greatly damaged irrigation systems between 1840–1770 BC and probably contributed to the collapse of the Middle Kingdom. The development of the New Kingdom (1570 BC) saw a number of innovations that were both geographically and organizationally significant. The introduction of the shaduf enabled far more control to be exercised over irrigation and water levels, moderating some of the effects of fluctuation. At the same time, the extensive colonization of the eastern side of the Nile Delta, by the new capital at Pi-Ramesse, led to a shift in the balance of population. The delta now became more densely populated than the valley. Yet by the twelfth century BC, food shortages resulting from the collapse of irrigation systems in the Nubian floodplain combined with invasions from the north to threaten the entire system. By 1070 BC, the New Kingdom had gone the way of earlier kingdoms (Butzer 1980: 517–23). As with Mesopotamian civilizations, material and ideological dependence on a highly prescribed ecosystem established a geographically emplaced rigidity at the heart of early Egyptian civilization, one that left it peculiarly vulnerable to the environmental fluxes of such a system. It need hardly surprise us to learn that whilst the Egyptian kingdoms displayed a capacity to overcome such crises and the discontinuities that resulted from them, even when compounded by other crises such as outbreaks of disease, they did so via geographical shifts of socio-political power.

Later Middle Eastern empires established bases of control that projected them far beyond the prescribed possibilities of riverine ecologies. Their increasing scale, though, brought new forms of instability. The Assyrian Empire grew out of a core area around Arbil and Nineveh during the early part of the first millennium BC, eventually forging a territory that spread from the Levant to Mesopotamia (see Fig. 3.2). Though dependent on strong military control, the focus of its power shifted between a number of core centres, being variously controlled from Nimrud, Khorsabad and Nineveh within the Assyrian heart of the middle Tigris valley. With the destruction of Nimrud and Nineveh in 612 BC, though, its rule collapsed. By the mid-sixth century BC, the core region of southern Mesopotamia had become absorbed into a still more extensive canopy of power formed by the Achaemenid Empire. With its hearth areas in the Zagros Mountains and Iranian Plateau, the Achaemenid Empire combined the agricultural wealth of the fertile plains that stretched along the Tigris and Euphrates valleys, the Levant, Anatolia, together with, on the east, the area stretching from the Iranian Plateau to the Indus valley and, on the north, as far as the valley of the Oxus. But again, with the defeat of its army in 331 BC and the sacking of its capital at Persepolis the following year by Alexander the Great, its vast sprawling empire fell apart. Within a century though, the heartland of the Achaemenid Empire – the Iranian Plateau, the Zagros Mountains and Mesopotamia – had been reconstituted as the Parthian Empire (240 BC–AD 226). This, in turn, gave way to the Sasanian Empire (AD 226–642), whose eastward and north-eastward expansion to the Indus and Amu Darya threatened to reinstate the form of the Achaemenid Empire.

Running through this succession of imperial forms that swept through the region were recurrent themes. As Watson has put it, what typified each new centre of imperial growth, including the Assyrians and Persians, was that they began as marcher societies or states (Watson 1992: 128). He does not see this as chance. Marcher societies, 'hardened on the dangerous frontiers or marches of a civilization whose centre had become softer and more refined', provided the impulse 'for tightening up a loosened system' (ibid.: 128). A further related factor, one noted by Mann in an argument directed towards the same end, was that marcher elites were 'lured by civilization and interested in acquiring it' (Mann 1986: 187). Other factors, such as the environmental differences posed by marcher areas and the way they fostered new solutions, can be seen as contributory. However, whilst marcher lords were often 'the bearers of interstitial surprise' (ibid.: 539), using new techniques and institutions to great effect, their success in forming eruptive empires tended to mix old and new. For this reason, Watson has suggested that from Assyrian Nineveh to Arab Baghdad, there was 'a remarkable continuity of imperial structure', with each new relocation of the imperial core utilizing 'the hegemonial structures that already knitted the system together' (ibid.: 128).

The Chinese Empire

Despite its different forms, the Chinese Empire has claims to far more long-term stability than most. Indeed, its traditional interpretation is based on a cyclical rather than progressive or serial model, with successive dynasties passing through a life-cycle of youth, maturity and old age, each reliving the same experience or processes (Lee 1965: 15–21; Reischauer 1965: 31–3). This life-cycle can be constituted in economic as much as political terms (ibid.: 33). Its outward impression of durability and sameness, though, hides interesting internal shifts in centres of power and dominance across time (see Fig. 3.3).

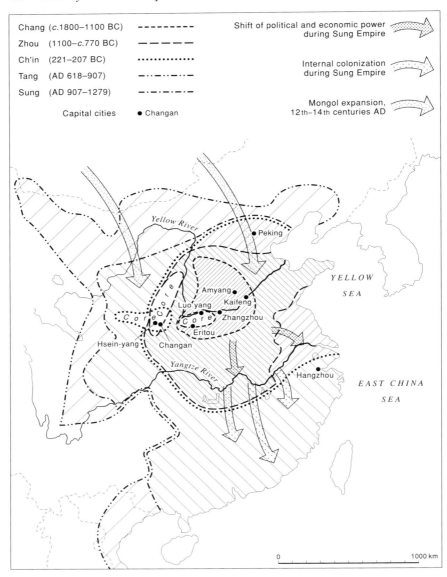

Fig. 3.3 Changing configurations of the Chinese Empire. The complex
reconfigurations of Chinese imperial structures over time make it difficult to
summarize all of them via a single map. The above map selects the Chang, Zhou,
Ch'in, Tang and Sung dynasties or empires and, through the mapping of their core
areas (in the case of first three), capitals and overall extent, tries to show the role
played by the periphery and by shifts in the *loci* of power. The Mongolian invasion

The first Chinese civilization, the Chang (*c.* 1800–1100 BC), developed on the rich alluvial soils that lay at the point where the Yellow River or Hwang Ho spread out over the North China Plain. As a river that changed course and even its sea outlet on a number of occasions during historic times, its potential for irrigating and fertilizing soils through controlled flooding was spread over a wide but hardly fixed area. Yet whilst these alluvial soils were important as an initial focus of settlement, the real thrust of growth came when the regular water supply provided by the Yellow River was extended out to the loess soils that covered the surrounding plains using an elaborate and highly regulated system of irrigation built and maintained using *corvée* labour (Lattimore 1962: 138–9).

Over its history from *c.* 1800 down to 1100 BC, the Chang civilization expanded out from its core on the western edge of the North China Plain to control parts of the wider region, laying the cultural foundations for subsequent Chinese civilizations (see Fig. 3.3). A sense of the internal shifts and adjustments that characterized its history, though, is indicated by its frequent changes of capital and political focus. Despite these successive readaptations of control, it had collapsed by *c.* 1100 BC under pressure from the eruptive expansion of the Zhou, a previously marginal or frontier group of nobles originating in the Wei-Ho basin of Shen-hsi province (Lattimore 1940: 355–6). The Zhou dynasty extended their sphere via a feudalized system of government, granting fiefdoms in return for military support. Underpinning their political expansion was a range of technical innovations that helped expand output, including the use of iron and, by the time of the later eastern Zhou, the use of controlled irrigation for bringing into cultivation many thousands of acres and, on the back of this increased output, the founding of many more cities (Tuan 1970: 52–4). No less important, between the phases of the western Zhou and the later eastern Zhou, there took place a shift in political power from the older but smaller states on the western side of the North China Plain to the later but larger states in the middle, from its early centres in the Wei-Ho valley to Lo-yang (Watson 1992: 866–7). By the eighth century BC, the dominance of the Zhou state and its dynasty had effectively broken down and for the next three or four centuries, the various 'warring' states competed for power. It was the western state of Ch'in, hitherto a marginal state based in the

Caption for 3.3 (*cont.*)
over the twelfth–fourteenth centuries AD established Peking as a capital, and paved the way for its use as the Chinese capital under the subsequent Ming Empire. Lattimore's point that early Chinese history was dominated by events along an east–west axis and later history by events along a north–south axis is brought out by the various shifts of capitals. Based on data drawn primarily from Seelan and Twitchell 1974: ix–xiv; Barraclough 1984: 62–3, 80–1, 126–7; *Past Worlds* 1988: 146–7, 192–5 and 262–3.

Wei-Ho valley around its capital at Hsein-yang, that ultimately gained the most from this disunity. Ch'in had been particularly successful in exploiting the potential of irrigation. Supported by an expansion of output and a new professional bureaucracy, it began to expand and to take political control of other Chan-kuo states and the fertile Ssu-ch'uan basin, initiating the extensive irrigation system of the Ch'eng-tu Plain. By the third century BC, it had briefly formed the first or Early Chinese Empire (Elvin 1973: 25).

The unity of empire established by the Ch'in was taken over and sustained by the Han dynasty (206 BC–AD 230). Abandoning the feudal character of the Zhou, the Han established a new capital at Changan and introduced major and sweeping changes in administration. Its rule, though, struggled to cope with the scalar challenge of empire and to balance the needs of the centre with the contrary needs of the periphery (Hall 1988: 22–3). Its use of a population census, development of paper for writing in place of silk and centralized control over key industries like iron and salt manufacture all manifest a system anxious to monitor and control from the centre. Yet the scale of its commitments and costs meant it also had to levy taxes from the provinces and to prevent the free peasantry from falling under the lordship of local estate owners and the very bureaucrats it empowered to run the provinces. The Han dynasty ended with a phase when this attempt to balance the needs of the centre with the periphery failed, and when the lack of unity between the different provinces was exploited in the north by barbarian invasions. After a period of instability and provincialism, the restoration of central control under the Tang dynasty (AD 618+) led to the formation of an even greater empire, with its extension westwards into central Asia.

Geographically, the main feature of the Tang period was the growth of agricultural production in the Yangtze valley. Traditionally, the greater proportion of the Chinese population, agricultural production and urban consumption was centred on the North China Plain. The disruption caused by the barbarian invasions in the north encouraged an expansion of production in the Yangtze valley. With the emergence of more settled conditions under the Tang dynasty, output in the Yangtze area grew even more rapidly, with a newly constructed canal built on a dramatic scale being used to transport grain and other foodstuffs northwards to the main markets. This greater exploitation of the Yangtze area began a process that was to continue under the following Sung dynasty (late tenth century AD+) with a shift not just of agricultural production but also of population and political power towards the south, though the Sung capital was at Kaifeng, at the northern end of the canal from the Yangtze to the Yellow River. Beyond the Yangtze, areas which had previously been little settled, like Fujian, became heavily settled, providing a rich frontier of new opportunity for the rapid expansion of Chinese society.

The Chinese Empire in all its various reincarnations from the Early Empire (221 BC–AD 311) through to the Later Empire (AD 960–1911) has been rightly regarded as amongst the most persistent of all empires. Like all extensive empires, its growth and consolidation depended on how it coped with two problems: the integration of new areas and the control of frontiers that, in time, became ever more distant and peopled by differences. In coping with these demands, it was probably no accident that for much of the first and for part of the second millennium AD, the Chinese empires were possibly the most innovative of world civilizations. Their progressive integration of new space followed a simple formula that took the form of 'the adding together of innumerable units, which in spite of local differences were essentially homogeneous, each consisting of a rural landscape watched over by a walled city' (Lattimore 1940: 41). As they matured, the management of these local segments, their exploitation via grain tributes, taxes, etc., on behalf of the core, passed into the hands of mandarins, professional bureaucrats whose loyalty lay with the centre rather than the provinces, an orientation ensured by a ban on their holding office in their province of origin, and whose evolving mastery of a codified written language and standardized record-keeping reflected the evolving needs of imperial control.

Like other expansive imperial structures, the more Chinese empires expanded, the more they had to deal with the problems posed by their frontiers, not so much those on the east and west as those on the south and north. The high ground to the west provided a natural frontier, whilst the lower plains and flood plains to the east were quickly and easily absorbed in the first four or five centuries of Ch'in expansion. The frontier regions that beckoned to the south and north presented different problems because they presented different environments. As colonization pushed southwards, and as the population expanded rapidly over the first millennium, they had to deal with environments that were tropical in character. However, the success with which existing systems of administration were extended southwards and the degree to which tropical crops, especially rice, were incorporated into labour-intensive husbandries meant that despite the scale of expansion involved, the southern frontier areas created few problems for the viability of the empire as a whole.

The same cannot be said of the frontiers that lay to the north and northwest. Here, the more physically broken and less fertile hill ground of the steppes did not favour the extension of agrarian systems based on intensively worked, river-fed irrigation systems. Local societies comprised highly mobile pastoral tribes. The territorial solution that was worked out over successive centuries was a complex one, based on broad inner and outer frontier zones. Over time, some frontier tribes underwent a process of sinicization, as Chinese communities were settled amongst them and as they, in response, became familiar with and adopted Chinese styles and skills. To a degree, these provided the empire with an effective buffer against less controllable tribes to the north and north-west. However, the frontier was not a stable affair. As Watson has also observed in the case of the Near Eastern empires, the fusion of acquired skills of administration and military technique from the core of the Chinese Empire with the hardened and warlike personality of the frontier tribesmen made for a powerful, uncontrollable mix. Indeed, phases of disruption, such as when barbarian armies overran northern China in the early fourth century AD, caused power structures to fragment and parts of the region to acquire a mixed ethnic character and, for a time, political independence (Elvin 1973: 42–53). The same chemistry of reaction between skills acquired from the core of the Chinese Empire and the ambitions of peripheral elites amongst the Tartars and Mongols led to more large-scale invasions across the northern frontier in the twelfth and thirteenth centuries AD. In fact, during the thirteenth and fourteenth centuries, China became part of a Mongol empire, a case of the periphery ruling the core. A notable symbol of its long-term impact on political geography lay in the way the capital of the Mongolian Empire, Peking, a capital positioned close to the frontier simply because the Mongols did not see the frontier in the same terms, became the capital city for the Ming Empire that eventually replaced the Mongol Empire in the fourteenth century (see Fig. 3.3).

Despite the persistence of the Chinese Empire as a whole, its survival has not been without significant internal shifts in the focus of political power. As Lattimore made clear in an early discussion of these 'shifts of gravity' as he called them, they were as much to do with the symbiotic interaction between frontier peoples and core areas as with the struggle between rival areas and dynasties within the heart of the Chinese Empire (Lattimore 1940: 350–61). Overall, Lattimore reduced the shifts to a simple switching of axes, with 'an east–west cleavage' being turned into a 'north–south opposition', and with the balance of control gradually shifting between the two poles of the north–south axis (ibid.: 311). A case can also be made out for seeing some of these shifts in the political centre of gravity as accompanied by fundamental shifts in the main centres of economic output, with successive governments helping to promote and sustain such shifts in output by directing new investment to public works in such areas (e.g. irrigation schemes, canals for transport) (Chi 1965: 65–8).

The Roman Empire

As a type of the species, the Roman Empire can be used to highlight a number of general themes as well as themes that are unique to itself. In keeping with other eruptions of imperial power, it developed from a peripheral position on the western fringe of what had been a well-established and innovative system of imperial power developed around the Greek city-states. From its origins as a city-state based on Rome *c.* 500 BC, its power was extended, via the direct extension of Roman territory, the creation of latin colonies or the forging of alliances with other states or tribes in the south, to cover most of Italy from about Pisa southwards by the mid-third century BC. By the mid-first century BC, it had established itself as a pan-Mediterranean power by securing control over Iberia, Corsica and Sardinia, Carthage, Macedonia (the Balkans and Greece), western Turkey and, belatedly, had taken control of northern Italy. In the following century, its runaway growth saw the conquest of Gaul, Cyrene in north Africa, further coastal areas of Turkey, Egypt and the Levant. By the first century AD, it had established secure control over southern Britain, pushed deeper into Germany, and secured Mauritania (north-west African coastal area), Dacia, Thracia and Asia Minor, and was about to expand into Armenia, Syria and the northern parts of Mesopotamia. Significantly, its expansion into western Europe had largely contended with locally organized tribal systems, whose initial resistance was crushed with great force, whereas the more developed and differentiated societies of the east had been incorporated by means of political alliances.

The empire tried to cope with the problems of both conquering and controlling its vast imperial space by the use of local resources to support its legionary forces and by incorporating local or marginal elites into its strategy of controlling new provinces. Yet the rapid growth and military character of the expansion had generated organizational problems even by the first century BC. A frequently voiced criticism is that its rapid and spectacular expansion was made without any corresponding adjustments to its organization, ruling a vast empire as an accessory to the constitutional forms of a small city-state, or Rome itself, forms that were narrowly defined and based on a senate dominated by its military elite (Nash 1987: 87). In its earlier phases at least, it made little attempt to devise 'an imperial administration that matched the dimensions of the empire' (Garnsey and Saller 1987: 20). Instead, it attempted to manage without a professional bureaucracy (Nash 1987: 88), whilst little effort was made to develop a uniform or systematic system of taxation, its tax policy being 'unstandardized, undersupervised' and undergoing 'little change' (Garnsey and Saller 1987: 21). Even its army, the mainstay of its control, relied on men drawn from its core areas in Italy until the end of the Republic (27 BC). Whatever the force of such criticisms, though, the empire initially coped well, albeit in what, from a pure administrative point of view, was a makeshift way.

Its expansion created huge logistical demands, with large standing armies having to be maintained and equipped along its greatly distended outer frontiers and regular flows of information maintained with the centre. In response to such problems, it skilfully exploited the resources placed at its disposal by the conquest of new areas, spreading the burden of empire across all parts (ibid.: 97). The extraction of resources in the form of tribute, though, had its costs. Strabo informs us that the reason why the Romans did not initially settle Britain and turn it into a province was because the high military cost of extracting tribute from troublesome British tribes, quite apart from the costs of maintaining legionary forces on its outer frontier against the Picts and the like, simply did not make it economic (Garnsey 1978: 238; Nash 1987: 100). Of course, there were areas where the input–output ratio of tribute or tax gathering was more clear-cut from the start. Egypt, for example, effectively became a major supplier of Rome after 31 BC, its access by water affording a cheaper supply than landlocked areas within Italy itself (Rickman 1980: 14 and 120). In fact, all the main grain supply areas of Rome were accessed by river or sea, including Latium and Umbria, Sicily, Sardinia and north Africa (Garnsey 1983: 119–20). Overall, the structure of exploitation within the empire amounted to more than just a simple core–periphery model. Predictably, Rome, and increasingly, Italy at large, represented a centre of consumption in terms of goods and wealth, a pattern well established by the late Republic. However, the various provinces served as more than just a vast area of exploitation for Rome itself. As Hopkins has argued, we can see them in terms of an inner and outer ring. Because of the armies positioned on the frontier, the outer ring that fronted the barbarian world actually formed a belt of consumption. The inner ring formed a 'relatively rich tax exporting area' (Hopkins 1980: 101), supplying wealth in the form of tribute, produce and goods both to the legionary armies of the outer ring as well as to Rome itself. The outer ring could rival the centre in other ways. The military power concentrated in frontier areas enabled powerful military commanders like Caesar to establish a base in the provinces and to use it as a base for their claims to power in the centre. The political struggles that had emerged in this way by the first century BC threatened the empire itself.

Renewed stability was secured for a time by the Augustan Revolution, which ended the Republic and initiated the age of empire. This sought to adjust the empire's organization to cope more effectively with its sheer size, acknowledging the simple fact that it now was an empire. Instead of a rule by Rome and its citizens in their interests alone, with its 'old, quite anachronistic political system' (Alföldy 1985: 93), it created an administrative structure system that began to address the needs of the system *in toto* (Levick 1985: 2–3). The administrative reforms of Augustus (27 BC–AD 14) recentralized the army, brought the rule of the provinces under his direct command and created a bureaucracy with responsibility for running the empire as an empire not as an adjunct to the politics and power struggles in Rome. These reforms effectively removed the running of the empire from domestic politics and created the conditions under which it could persist for longer (Doyle 1986: 95). Though some have talked about its excessive centralization and its use of 'an immense volume of clerical labour' (Jones 1964: 1056), they have conceded that even after the Augustan reforms and the establishment of the empire, its bureaucracy was 'not vast in relation to the size of the empire' (ibid.: 1057). In fact, others have gone further and concluded that it was probably an undermanaged system given its vast scale and complexity when compared to the

Chinese Empire (Garnsey and Saller 1987: 40; see also, Hall 1988: 20), with an estimate of around 30,000 regular officials administering a population of *c*. 50–60 million people (Jones 1964: 1057).

By the time of Diocletian's rule from AD 284 onwards, the temporary stability gained by the Augustan Revolution had already started to crumble. The essentially civilian government introduced by the latter had been threatened when, following new difficulties around the borders of the empire, a number of military governors tried to assert their independence. The Diocletian response to these problems was again to restructure the empire by dividing its control between two rulers each with a deputy, and to sub-divide the empire into four large prefectures that were further sub-divided into dioceses. Though this attempt to systematize power overcame some of the problems faced by the empire, it was not a permanent solution. Very soon, the domain of the two rulers became split into an eastern and western empire, the eastern acquiring a new capital at Byzantium or Constantinople in AD 330. As outlying parts of the empire were overrun, especially by Goths, Vandals and Picts in the west and north, but also, by Slavs and Arabs in the east, and after Rome itself was sacked in the fifth century AD, the Diocletian division of the empire became crucial to developments. The western half collapsed. By comparison, the eastern empire, though greatly reduced, managed to survive.

Much has been written about the reasons for the decline of the Roman Empire. For some, its problems were structural, and rooted in the plain and simple fact that it never really came to terms with the scalar problems of being an empire despite the Augustan and Diocletian reforms. From its beginning to the late Empire or principate, said Heitland, 'Rome never developed a political organ capable of continuous action or peaceful reform' and failed to appreciate the extent to which the increasing size of its political reach did not simply pose 'differences in scale' but 'differences in kind' (Heitland 1962: 58 and 64). To compound matters, it did not make up for these weaknesses of administration by creating a shared sense of political identity amongst those it conquered, with most of those ruled being 'passive' about the empire or having 'little experience or conception of Rome' (Garnsey and Saller 1987: 203). More serious, whilst it was resourceful in exploiting new areas of conquest, uplifting vast supplies of food, raw materials and manufactured goods for its core elites and legionary forces, it was not able to secure these supplies in the long term. By the second and third centuries AD, sharp falls in population caused by disease (Walbank 1969: 61) and the extensive abandonment of land, *agri deserti*, estimated at over half a million hectares (Jones 1966: 366; for a different view, see Lewit 1991), together with a general decline in trade, greatly weakened the capacity of the late empire to support itself. Not surprisingly, these problems affected the empire's tax base, reducing what could be raised as tax but not what was needed. For many, the empire's tax burden and the extent to which it surcharged provincial production may not have been the sole cause of its political decline, but it was the most significant single factor. It was a 'crushing burden' concluded Heitland (1962: 64), and 'probably a root cause of the economic decline of the empire' (Jones 1966: 366), reaching 25 per cent of output in the late empire (Jones 1964: 1057). Inevitably the exact figure varied between areas. In Egypt, farmers paid as much as 50 per cent of their crop as tax or tribute, whilst, by comparison, those in Italy paid little in the way of direct taxation prior to Diocletian (Garnsey 1978: 241). When coupled with the collapses of population (itself a source of *agri deserti*) and the growing barbarian devastation of the western and northern frontier areas, leading eventually to their disruption of areas and output deep within what had been *pax Romana*, the empire's weaknesses of structure and maintenance became overwhelming.

Subsequently, the two parts of the empire experienced contrasting fortunes. The western half collapsed completely. However, what had been peripheral areas either just beyond its frontier (e.g. Ottonian Saxony) or within its outer provinces (e.g. Alfred's Wessex) eventually became the foci for the development of a new and innovative political form in the shape of the feudal state (Rokkan 1975: 576). The richer, more fertile eastern half, meanwhile, survived, being reborn as Byzantium. Having retreated before vigorous Arab expansion in the east and in Africa and before the southward expansion of the Slavs, Byzantium came under the rule of a strong, military-based Macedonian dynasty in the ninth century AD. For a time, its territorial spread over the eastern Mediterranean was restored. Yet no sooner had the initative been seized than it was lost. Its army shifted from being based on the free peasantry to being organized around mercenaries. More damaging, from having an open, bureaucratic system, provincial families who secured important positions over law and taxation fought successively to make their position hereditary. As with Flannery's shift from institutions that are system-serving to institutions that are self-serving, the effect was to close down the opportunities for further change and responsiveness (Flannery 1976: 106–7).

From commercial empires to modern world system: the post-1500 world

As mentioned earlier, since the idea of a capitalist world system was first defined by Wallerstein (1974), the concept has been much exercised. For Wallerstein, world systems or empires had existed prior to 1500, but what emerged after 1500 was a world system *differentiae specificae* (Wallerstein 1993: 292–6). For some, his case is weakened by the fact that not only did world systems exist before 1500, but they were also capitalist, being based on market-centred trade and capital accumulation (Gills 1993: 118–19). In reply, Wallerstein would point to the geographical scale of what emerged after 1500, its long-distance trade in bulk commodities rather than just luxuries, and an 'axial' division of labour between different parts of the system as features which give identity to the world system after 1500. Like Wallerstein, I would plead the case for seeing what developed after 1500 as distinctive when compared to earlier world systems or empires. Admittedly, part of the case has been qualified by work on pre-1500 world systems and the growing evidence for the role played in their development by market activity, long-distance trade and capital accumulation (see, for example, Silver 1983; Garnsey 1983: 126–7). However, there are still key aspects of the post-1500 world system that were not pre-figured in earlier systems. One particular point, a point which is well made in Polanyi's work and which – despite continuing attempts to describe his substantivist ideas as flawed – is still valid, is the extent to which pre-1500 markets were free or self-regulating. At issue here is more than just the question of how markets were regulated on the ground. Even in traditional societies, short-term fluxes in food supplies especially, could greatly affect prices, however much authorities tried to regulate them within socially acceptable limits. What mattered, though, is the broader ideological context in which markets and trade operated and the ends which their activities were allowed to serve. Polanyi argued that early trade and markets were socially and politically embedded (Polanyi 1957: 57–75; Polanyi 1968: 158–73). In the process, they served to ensure an allocation of property and resources that underpinned prevailing socio-political structures and their elites (Gilpin 1981: 112). They were allowed to be competitive only so long as their operation did not threaten the roles and values given by these structures (see, for example, Garnsey 1983: 126–7). What was different about the markets which emerged after 1500, though in a way that developed more slowly than Wallerstein would have us believe, is the extent to which markets and their capital flows became more autonomous, more self-regulating, as free markets became the prime means by which capital was accumulated. Such markets are still socially embedded in the sense that some social groups still gain from their activity, but increasingly, they operate across – and beyond – political structures, and become legitimated in their own terms (i.e. what maximizes accumulation *via the market*), with players and winners being defined, first and foremost, as those who risk capital rather than simply as those who have a monopoly of power and coercion. It is this growing autonomy of market values, with governments actively taking steps to make their workings more efficient not more encumbered (Jones 1988: 135), that distinguishes the modern capitalist world system from earlier systems.

Its autonomy, combined with the greater openness of rules that govern free markets, has meant that the modern world system is not a stable system, but constantly seeks to expand its arena of operation, both geographically and sectorally. It also seeks to maximize the difference between the markets in which goods and labour are bought and those in which they are sold. To this end, it is constantly engaging and disengaging investments. For this reason also, it purposefully generates change and instability *within itself*, as capital seeks out new areas of investment at the margins of the system and, in doing so, continually disentangles part of itself via profits from older areas. Again, we can see this purposeful generation of instability as marking a qualitative distinction with earlier world systems, which – internally – were purposefully organized for stability even if this was not achieved. These processes of growth and change under the modern capitalist world system are depicted in Fig. 3.4.

In its initial transitional phase, though, the modern world capitalist system was put together and sustained by military force in a way that hardly differs from the way pre-1500 systems were created through the growth of powerful, aggressive imperial structures. To put this another way, whilst commercial interests prompted the early voyages of 'discovery', actually seizing and securing the commercial opportunities released by them required the coercive power that only governments could muster. Whether through formal trading alliances with local rulers or through military conquests and claims to sovereignty over new-found areas, core societies – notably those of Portugal, Spain, Holland, Britain and France – established a framework of monopolies within which privileged trading companies could develop trade. Very quickly, reciprocal flows developed, with bullion as much as agricultural products and raw materials flowing in one direction and new colonists and, eventually, manufactured products in the reverse. In reality, of course, such flows were not organized on a simple two-way basis, but were complicated by a vast human movement of slaves from one part of the periphery to another, enabling colonial producers to reap the low-cost benefits of highly coercive and exploitive systems.

Over time, trade broke free from the strait-jacket of state-granted monopolies and the handful of trading companies that enjoyed them. However, there was still a phase during which it continued to follow the flag, with a welter of trading restrictions being used by different core countries to ensure that trade between core states and their colonies was largely restricted to their own merchants and fleets. By the mid-nineteenth century, a more significant change had started to unfold with the shift into a more consolidated international system of free trade. As free trade came more and more to dominate the world economy, to play a determinate role in shaping the ties that bound it together, so an increasingly integrated global economy engendered a world capitalist system driven by the needs of capital. In other words, the economic links have become more autonomous not in the sense that they are no longer set in a social or political context but in the sense that economic considerations have become the primary organizing force behind the world system. Whilst Wallerstein would date the inception of this world system to his long sixteenth century, the early years of the world system only promised the kinds of integration which lie at the heart of his definition. Without taking this essential meaning away from the concept, there is a strong case for arguing that its emergence as a functionally integrated system, one capable of allocating economic roles to different sectors of the system, needs to be delayed until the

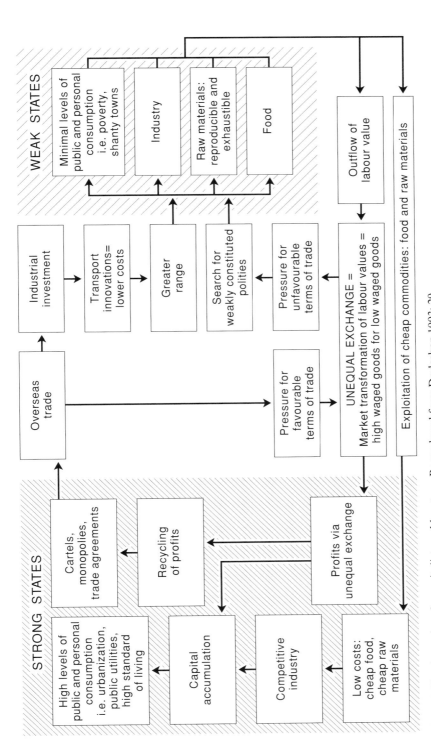

Fig. 3.4 The dynamics of a capitalist world system. Reproduced from Dodgshon 1993: 29.

mid–late nineteenth century (Bairoch 1974: 560–1; O'Brien 1982: 4; Dodgshon 1993: 26–41).

The potential for this new kind of commercially driven empire was first established during the 1490s with the Portuguese and Spanish voyages of exploration. These voyages have a similar background. Faced with growing Turkish control over the eastern Mediterranean and the problems that it created for the overland route to India, China and the East Indies and for the trading of sugar from the Levant, Italian merchants – those in Genoa and Venice especially – were anxious to find new ways of reaching the East Indies so as to maintain flows of spices and sugar and to maintain their grip on the trade. In fact, it was Genoan merchants who helped to fund the voyages made by Columbus in 1492, supposedly to find a new way to the East Indies but finding the West Indies instead. The early Portuguese voyages were a similarly profitable alliance with Italian venture capital and were prompted by the same goals. The Portuguese no less than the Italians were equally interested in the possibility of opening up new routes for the spice trade and, even at this early stage, in obtaining supplies of slaves. By the 1480s, they had reached the Gold Coast and by 1488, Dias had rounded the Cape of Good Hope. When this information was combined with that provided by an exploratory voyage down the east coast of Africa from Ethiopia, it provided the Portuguese with the knowledge needed for a sea route to the East Indies. We also need to keep in mind Abu-Lughod's point that the world system previously structured from China to Europe, her thirteenth-century world system of the east (Abu-Lughod 1989), had collapsed by the time the Portuguese reached the Indian Ocean and, as a consequence, the ports bordering the Ocean were not organized in a way that could resist the Portuguese intrusion (Abu-Lughod 1993: 288).

The empires that developed from this surge of European exploration shared a number of characteristics. First, in character, they were essentially maritime empires, binding together areas, ports and preferential trade agreements that were discontinuously spread over vast oceanic distances. However, there is this distinction to be drawn. The earliest maritime empires, or those of Spain and Portugal, were founded as much on land-based military conquests and plunder as on trade, especially in Central and South America, whilst the later empires, notably those of Holland and Britain, were driven more by trade and systematic colonization supported by, rather than led by, military coercion (Meinig 1986: 43–51). Second, whatever their character, maritime empires involved large-scale social disruptions and movements, with native populations in peripheral regions being savagely decimated by warfare and disease, large numbers of others being mobilized and moved to sustain elaborate systems of slavery, and new populations migrating from the core to peripheral areas. Third, economic exploitation and unequal terms of trade became a primary feature of the commercial links between core and periphery and an increasingly important *raison d'être* of how the world system at large unfolded. Fourth, the structure of commercial empires differed from wholly land-based empires. The latter were relatively tidy, non-overlapping affairs, their respective interests separated by frontier zones but rarely by areas in which interests were shared. By contrast, sea-based trading empires were more numerous and less easily disentangled. Areas like the West Indies became closely divided between a number of rival empires, so that state navies had to be continually used to fend off rival claims as much as to protect against piracy. In the first instance, the dense network of criss-crossing movements that gave birth to the Atlantic economy comprised a number of empires, not just one. Only with the emergence of larger, more integrated systems, and with the shift from closed systems of regulated trade to open, freer forms did there emerge a more aggregated, inclusive system, the so-called modern world system. Only when this point was reached can we accept that we are dealing with the history of a system whose whole was greater than its parts.

Yet despite its initially compartmentalized form, the incipient modern world system did have a dominant sector and port whose trade appeared more vigorous and innovative and which, by assembling a greater range of goods, became the system's base point, the point of maximum availability and information, from which European prices generally for spices, tobacco, etc., were derived. From its incipient formation as an extended system of trade back in the sixteenth century, this dominant centre or core of the modern world system, like earlier world systems (Gills and Frank 1993a: 102–3), underwent what Braudel called a progressive recentring (Braudel 1977: 85). Generally speaking, each major shift of dominance between the various sectors or metropole centres was sustained by a substantial shift in capital and commercial advantage. Also underpinning these shifts in advantage were political changes, changes in the opportunities being exploited, both as regards areas and products, in transport technology, in the organization of trade or in the institutional structures through which it was operated (Hugill 1993). In effect, the changing geography of the various commercial empires, like that of the world system that gradually overwhelmed them, was part of the tactics of change, the means by which radical changes could be introduced and advantages gained.

Reviewing this succession of change in more detail, European trade over the late medieval period had been dominated first by Venice, then by Genoa. Each flowered, profited, then stagnated. Each exploited, and gained from, a particular set of opportunities and advantages. With the dramatic broadening of the European horizon over the fifteenth century, and particularly with the Portuguese exploration of the African coast in the 1480s and Columbus's Atlantic voyage in 1492, Spain and Portugal moved quickly to exploit the possibilities of empire. Between them, they began the shift from a trade organized around the Mediterranean economy to one organized around the

Atlantic economy. As countries located at the point where the Mediterranean gave way to the Atlantic, and at the corner of Europe closest to the west African coast and to Central and South America, they were best placed to channel the needs, skills and finance of earlier trading systems into the new opportunities of the Atlantic economy. Indeed, although they were associated with the use of new or better technologies (the carvel, galleon and navigational skills), the real advantage exploited by Spain and Portugal was locational, being best placed to project the capital and energies of the Mediterranean out across the Atlantic.

Spanish efforts were concentrated in Central and South America. By the treaty of Tordesillas (1494), they conceded prior claim over potential colonies in eastern South America to Portugal, the conquistadores themselves eventually seizing control over Mexico, Venezuela, Bolivia, Peru and parts of coastal Chile. As well as the vast amount of bullion plundered from native societies, they developed successful but short-lived silver mines (e.g. Potosi in Bolivia). For a time, Seville, and the Spanish economy at large, received huge injections of wealth, injections that are permanently recorded in the casas built by conquistadores at places like Trujillo. The growth of the Spanish Empire, though, was insecurely founded. It relied more on what it could gain by military conquest and by plunder than on new trading opportunities or through innovations in the organization of trade. It has been said, perceptively so, that its approach to empire built on their spirit of internal *reconquista*, with the first overseas exploration occurring soon after Moorish political power in Spain had finally collapsed (Meinig 1986: 43–51). Certainly, its exploitation of new colonies redirected some of the tensions within Spanish society and provided an abundance of what might be called free-floating resources, but it was not a sustainable change. Much was consumed conspicuously and the flood of bullion led to inflation. Indeed, as Ringrose has pointed out, the scale of Madrid's consumption had the effect not of stimulating the Castillian economy but of draining it (Ringrose 1989: 57–80).

The Portuguese Empire pieced together from the late fifteenth century onwards displayed more variety and innovation but it did not lead to sustained economic or social change. The trade which it developed had three components to it. First, it established flows of gold from colonies opened up along the west and south-western coasts of Africa. Second, using its ports and staging posts along the coasts of both west and south-east Africa, it opened up a valuable maritime trade in spices with the East Indies, organizing the trade from its colony at Goa. Third, using slaves drawn from Africa, it opened up sugar plantations in Bahia, the nucleus of its Brazilian colony. Gold, spices and sugar were longstanding commodities of trade. Portuguese success involved the opening of new source areas and new trading routes for them, with much overland trade to Asia and west Africa now being routed via the Cape and/or the west African coast. Reinforcing this success were a series of technical innovations, including the introduction of the galleon, a vessel more suited to oceanic traffic, and improvements in navigation and cartography. Yet as with Spain, the trading advantages were not cumulative or sustained, nor did they trigger internal change. Part of the problem was that whilst the opening up of new areas and trading routes was outwardly under the political control of the Portuguese, a great deal of the financing and commercial organization of trade was actually in the hands of Genoese merchants. Furthermore, the Portuguese struggled to retain control over their new trading opportunities. In the East Indies especially, their control was undermined by the Dutch, whilst even in Europe, it was Dutch merchants and vessels who carried the sugar and spices on to Holland. Having promised so much, the trading advantage gained by *c.* 1500 had faded by the mid-sixteenth century as control shifted northwards to Antwerp and Amsterdam. Indeed, having accumulated so much at the start of the sixteenth century, the Portuguese crown was bankrupt by 1560 (Scammell 1981: 291).

Antwerp had emerged as the dominant centre of the Atlantic economy by the 1540s. Its growth owed much to the way in which its merchants and banks had penetrated Portuguese and Spanish trading systems at their European end and had provided not only finance and organization but had developed a powerful role in linking them to the wider European market. More significantly, it gained considerable advantage from the way in which it combined trading flows from the Portuguese and Spanish empires with the trade that flowed out of central Europe via the Rhine and with an expanded trade to and from the Baltic. To a degree, the elaborate trading system built up around Antwerp was the first that hinted at the modern world system, with the different trading systems being integrated together into a network that transcended political systems and with the needs of trade driving its direction of development. Its opportunism was reinforced by far-reaching innovations in the organization of trade. It was at Antwerp in the mid-sixteenth century that the first real stock exchange developed, enabling information about commodities and their prices to be pooled and, almost as a consequence, it was at Antwerp that we find the first signs of speculative dealings in stocks and commodities. These changes in practice were matched by radical developments in financing, with ventures being funded by capital drawn from beyond the business partnerships actually engaged in the venture, and by the innovatory use of discount bonds and low-interest loans. Yet for all its exploitation of these new opportunities, the success of Antwerp was not sustained. By the end of the sixteenth century, it too had lost its role as organizing centre for the Atlantic economy to Amsterdam.

In some respects, the success of Amsterdam involved regressive changes. The growth of its trading empire depended heavily on the use of its military power to secure trading advantage and to protect its merchant fleet, together with the creation of monopoly conditions for its merchants in particular areas and commodities. In other words, there was a degree of political closure about the trading empire which it put together. This use of power and monopoly to extend trade through the agency of the Dutch East India Company was especially apparent in its expansion of trade with the East Indies, taking over not only Portuguese staging posts on the African coast but driving them out of key Asian ports like Cochin, Colombo, Malacca and Macao, and extending the spice trade into new areas like Indonesia where it established factories for cloth production. Its penetration of the West Indies was less extensive but still sufficient to give it a primary role in the sugar trade to Europe, whilst its capture of Portuguese trading stations along the African coast gave it control over the slave trade to the West Indies. Like Antwerp before it, Amsterdam's commercial hegemony also rested on its success in taking a commanding control over the lucrative network of Baltic trade and in expanding it considerably with regard to the movement of basic goods like grain and timber and luxuries like fur. Having secured its position at the centre of different networks of world trade – the East and West Indies, Africa, the Baltic and the Mediterranean – Amsterdam developed a hugely successful role as a transhipment port, drawing in a vast range of commodities, around 800 in one estimate, from a wide range of sources and organizing their distribution to an equally wide range of markets. Its control over shipping was matched by its control over information about trading opportunities (Smith 1985: 985–1005). The institutional and technical side of trade responded to this scalar expansion of trade and information. Stock companies were evolved that had a life beyond the individual venture (e.g. the VOC), standardized techniques of banking and handling goods were developed and the fluyt, a large, bulk-carrying cargo ship, was introduced (Steensgaard 1982: 235–57).

By the early eighteenth century, the focus of world trade had started to shift yet again. This time it stemmed from the success of France and Britain in building rival systems of world trade. For much of the eighteenth century, it was an open question who would succeed in wresting the baton of control over the world economy from the Dutch. Both had secured bases in the East and West Indian trade by the mid-seventeenth century. In the former, the English established secure stations at places like Madras and Tellicherry and in the latter, on Jamaica. The French, meanwhile, established stations and ports in the former at Pondicherry and in the latter on Saint Domingue. Under cover of France's growing naval power, French merchants secured more and more trade in both sectors. Significant for their long-term expansion was the fact that both Britain and France had established trading ports and colonies in North America, the English concentrating on the western seaboard of America and the French on the Great Lakes and inland along south-flowing rivers like the Missouri. In the struggle between them over the eighteenth century, it was the English that gained most. They established a dominance in the trade to America, a dominance that survived the American

War of Independence. They exploited to the full the triangular trade between Africa, the West Indies/America and Europe. In India, they broadened their influence to embrace most of the eastern coast and most of the middle and lower Ganges by the end of the eighteenth century. As with Dutch commercial hegemony, English control involved the combination of military force coupled with the regulation of trade through the granting of monopolies to home-based stock companies, merchants and vessels, and the use of high protective duties combined with export subsidies. However, it added a powerful new factor to its growing dominance of the world economy by the late eighteenth century. Not only did it profit from the wealth provided by trade and the new areas of production opened up for resources like sugar, tobacco, spices, cloth and fur, but it increasingly used colonial areas as a source of raw materials for industrial production (e.g. raw cotton) and as a market for its manufacturing sector. In other words, it not only raised the level of interaction between core and periphery, but it also created a more reciprocally based and intensified relationship. The degree of interdependence between Europe and the colonies at this point, though, must not be over-estimated. Certainly, huge profits were being made out of the Atlantic trade, as the exceptional concentrations of wealth in ports linked to it, like Bristol, Liverpool and Glasgow, testify to. Some have gone further and presented a case for seeing the profits made out of trade as exceeding those generated from home manufacturing (Blaut 1992: esp. 36–53). However, when we look at figures for British trade over the late eighteenth and early nineteenth centuries, they suggest that it was still far from being functionally *dependent* on foreign trade. Such dependence and, with it, the realities of a mature world system in the strict sense meant by Wallerstein, did not really come until the mid–late nineteenth century when rising levels of urbanization coupled with the falling price of imported grain quickly brought about a dependence on imported foodstuffs and when a sharp scalar increase in industrial output brought about by larger, more mechanized factories created a demand for new and larger overseas markets.

Altogether, the growing British dominance within the world economy was sustained by its heavy involvement in a number of far-reaching changes and adjustments that reworked the geography of the world economy. First, it was instrumental in opening up more distant areas for the European economy like Australia and New Zealand, securing them politically and then exploiting them socio-culturally and economically though colonization and trade.

Second, it played a substantial role in opening up the interior of areas long exploited only on a marginal or fringe basis. Everywhere, from the areas around the Baltic to the American mid-west, the exploitation of resources by the world system was pushed inwards into previously more landlocked areas, the process being driven by the growing demands of a European urban system whose base point for determining the scale of demand, organizing information and setting prices was now London. In the case of the American mid-west, this opening up of the interior was underpinned by major economic and socio-cultural changes as vast new areas were settled, highly efficient systems of land-holding laid out, new transport systems established and their farm economies drawn firmly into the wider world economy. In the areas around the southern and eastern edges of the Baltic, it was more a case of transforming the old rather than the new, with the economy of long-established aristocratic estates close to major rivers like the Elbe, Oder and Volga becoming reorientated through an increasingly coercive use of peasant labour, the so-called second serfdom, towards a commercial grain economy. In the case of Africa, too, it was a political as well as an economic process. Long exploited on a peripheral basis, the nineteenth century saw a scramble to secure political control over its interior regions, a scramble which left virtually the whole of the continent under some form of foreign or colonial control and which clearly broadened the scope and range of opportunities for subsequent commercial exploitation. Left with a command of territory in all corners of Africa, Britain secured more advantages than most from this scramble. Taking a broader view of the changes in Africa, they can be seen as part of a changing geography of colonialism as first America, then South America were politically decolonized but new areas, like Africa, large parts of the Indian sub-continent and south-east Asia, were subjected to European political control in their place. Yet for Britain, the idea that trade follows the flag had already ceased to be a force behind empire-building by the early nineteenth century. Aware of how much it gained from areas or sectors in which its merchants had no protection as from areas in which they had protection, or from its informal empire as much as from its formal empire, the balance of argument was swinging in favour of free trade by the 1840s. Its risk-taking in this direction paved the way for the great surge in cheap food imports which characterized British trade over the second half of the nineteenth century and which did so much to underpin its shift to an urban industrial economy.

Third, by the mid-nineteenth century, British trade gained considerable benefit from its primary role in overcoming the frictions and costs of distance. Its introduction of larger, quicker vessels, refrigeration facilities and ultimately iron-clad ships all aided long-distance trade, bringing the costs of shipping over long distances down dramatically (Harley 1988: 851–76). Likewise, Britain's development of railways, and its success in developing the railways of other continents such as those of South America and India, contributed greatly to the opening up of landlocked areas for global trade.

Fourth, Britain changed the nature of core–periphery relationships by initiating a huge export of investment capital. Indeed, down to 1914, when the hegemony of Britain was already starting to slip away, the London stock exchange was largely concerned with the export of capital into overseas investments and securities. This outward flow led to a substantial return in interest and profit but at the cost of starving home-based industry of investment at a time when its needs had long passed beyond the bounds of simple business partnerships (Ingham 1984).

Even as Britain finally secured its hegemony over world trade with its shift to free trade in the 1840s, changes were in place that eventually undermined its position and that of north-west Europe as the dominant core of the world system. Though Germany threatened to establish hegemonic control over the modern world system (Hugill 1993: 14), it was the United States that actually did so. Starting in the 1840s, the latter began a process of large-scale industrialization, with a shift from home to factory production, that lifted it into a lead position within the world system by the end of the century. The parameters of change have been well summarized by Agnew. By the 1870s, the challenge of connecting its potentially vast space economy had been met not only with the development of river- and water-borne systems, but by the rapid build-up of investment in a railroad system that surpassed that of any continent in the world. By the 1890s, its industrial output had become greater than that of any other country and American companies had started to invest overseas. By the First World War, it had exploited the opportunities of free trade to establish itself as the major source of world trade, whilst by the Second World War, the New York stock exchange had become the leading financial centre in the world, handling a greater volume of investment and information than any other centre. As with Britain in the late nineteenth century though, an increasing amount of its investment capital by the 1960s was starting to seek opportunities overseas in areas which formed part of America's informal empire (Agnew 1987: 85 and 146–9).

Unlike earlier centres of the world economy, whose growth had developed around fairly mature socio-political systems and had been shaped by that fact, the paramount feature about American growth was its high level of free-floating resources, or resources that were not locked tightly into prior uses. Combined with a massive influx of European immigrants and their skills, especially over the nineteenth century, economic growth and with it, institutional change, was exceptionally rapid and far-reaching. Its irruptive character was shaped in a number of ways. First, its history follows the classic course of a periphery turned into a core. Meinig explores this in his *Shaping of America*, stressing the role which provincial centres play within the periphery of all empires and not just America, as opposed to the role marginal areas of the periphery play within the collapse of empires as marginal elites and those trained to administer the margins turn against the centre (Meinig 1986: 373–4). In fact, Meinig sees America itself as becoming constituted as 'an empire in very basic ways' (Meinig 1993: 192–3 and 516). A second factor that was influential in its history was the role of the frontier, a seemingly limitless space of opportunity that beckoned colonists westwards for nearly 400 years. Turner's suggestion that the frontier and its intensifying settlement evolved through stages that mirrored the history of mankind has now been rejected. However, the idea that the frontier created an American personality, a person that was no longer European, finds echoes in what has been argued for marcher or frontier mentalities elsewhere (MacKay 1977: 2; Burns 1989: 315). In the case of America, though, the sheer scale and physical character of the frontier coupled with its colonization through communities that maintained their cultural identity if not the socio-political context from which they had come meant that whilst there was but one species of American frontiersman and woman, there were many different sub-species (Conzen 1990: 221–48). No less pertinent to the making of American character was the fact that many frontier areas, especially those planned out over the eighteenth and nineteenth centuries, created farm units that were far more efficient than their European counterparts. Once linked by river and rail to the seaboard and by faster shipping to the European markets, their grain output had a devastating effect on the very European countrysides that had sent colonists to the American midwest in the first place. Third, once its rapid expansion had set in over the mid–late nineteenth century, the organization and technical character of American industry soon enabled it to displace Britain as the workshop of the world. Chandler has argued that between the mid-nineteenth century and the early twentieth century, the American industrial enterprise passed through three critical stages, as it moved from single product and single-function units, to single product and multi-function units, finally evolving into the multi-product and multi-function enterprise (Chandler 1988: 13). The last-mentioned demanded administrative co-ordination, monitoring and allocation (ibid.: 400) and, by the early decades of the twentieth century, had provided the institutional basis for the development of assembly-line working and Fordism. In part, Chandler sees such attributes of organization as deriving from the sheer scale of geographical space over which American industry operated. As a form of institutional change, it originated in the management of the early American railroads whose sheer scale of spread and operation created new demands in terms of co-ordination and monitoring and, from the 1850s, led to new forms of business organization (ibid.: 400). In a sense, the challenge of American railroads matched that of other large-scale systems, like the Roman and Chinese empires, whose achievements in connecting space were similarly fundamental to their success as large-scale systems.

I have tried in this chapter to draw out some of the ways in which empires and world systems change. Though most were energized by various forms of technical or organizational innovation and change, in addition to the crude exercise of power, the advantage conferred by such innovations was rarely sustained in the long term. Indeed, what stands out about the geography of empires and world systems is the degree to which it has been subject to regular remapping. Even the strongest of systems cannot claim stability, being open to internal shifts and/or eventual eclipse. As Rowlands has pointed out, these regular or cyclical shifts within core areas and between different hegemonies introduces a geographical component to the problem of change, one that he labels as 'a theory of shifts of centres in space' (Rowlands 1987: 10). Furthermore, there has been a tendency for the cores or hegemons of new empires or systems to grow out of the peripheries of earlier systems. What this means is that the problem of change may not only be something that experiences a geographical shift as part of its unfolding, but it might also be tied up with what can be called the 'situatedness' of particular societies and their latent opportunities for change or adjustment. The problem needs to be seen as two-sided. Despite their dominant position, cores and their hierarchies acquired conditions which prevented them from maintaining the upper hand, just as peripheral areas often acquired the means and conditions under which they could successfully assert their control over a new and more dominant system. The fact that such peripheral areas often 'free-loaded' on the institutional and technical advantages already provided by their former core, as well as adding their own institutional and technical innovations, suggests that the two sides of the problem sometimes had a symbiotic relationship to each other.

4

The experience of change: states and regions

Much of the detailed analytical work dealing with geographical change has naturally focused on change within the framework of the state. In reviewing some of the themes provided by such work, my intention is not to show that such change takes place. That would be to repeat what is patently obvious. Rather is it to show that there are other neglected dimensions to the argument. Despite the changes that have helped transform them, there is also a case for arguing that states have experienced powerful inertias or rigidities that have acted to retard, deflect or refract these changes geographically. As with the kinds of change outlined in the previous chapter, these inertias and rigidities force us to look beyond the pressures working positively for change and to incorporate the constraints acting on change into the analysis. What matters for the geographer is that many of these constraints are geographically specific. Indeed, a great deal of socially produced space is potentially inertial and therefore a source of constraint on further change. Once this is appreciated, we can more easily see how change is not about the unfolding of some limitless potential in a frictionless space but about how the pressures and forces working for change are constantly having to negotiate around these geographically emplaced constraints. To put this another way, change is about a potential that is continually being remapped through the social construction of different geographies.

The varieties of change that have worked to transform state systems since their first emergence provide ample illustration of this need for societal change to work around geographically emplaced constraints. Altogether, I propose to look at three different dimensions of the problem in so far as it affects states and their component regions. First, there is the question of how states have developed structurally, that is, as territorialized systems of political control. Second, there is the question of how they coped with all the changes that can be bundled together under the heading of modernization as a cultural process. Third, there is the question of modernization as an economic process and, in particular, how states have been affected by the transformative power of industrial and market capitalism.

States and the geographical emplacement of power

Compared with the chiefdoms and tribal kingships which they replaced, early European state systems represented fixed, territorialized systems of power. Discussions of the early feudal states that emerged in western Europe over the ninth and tenth centuries stress the extent to which rights and jurisdictions now became fixed in space, anchored to definite territories and blocks of land. Whereas earlier societies had established identity through the kin-ties and alliances that developed around chiefs and tribal kings, so that territory was a consequence of its social content and could expand or contract with the success or failure of those who gave it identity, the feudal state began with the abstract concept of a feudal domain within which the king or ruler was lord of all men and all land. Territory, as the domain of state rulers, now became the primary reference of political identity. As Sahlins observed when discussing this transition as a general societal process, we move from a world in which territory is identified through society to one in which society is identified through territory (Sahlins 1968: 5). Of course, like all such generalizations, there are qualifications to be made here. Some social theorists would argue that states, early forms no less than later examples, are best seen as structured around overlapping networks of power and concentrated around particular nodes of power, not spaces. At risk of trying to separate out the interests of the geographer from those of the social theorist, such a view ignores the fact that precisely what distinguishes the early state from previous political forms is the fact that the jurisdiction of state rulers was now enframed in a continuous space or territory as a pre-condition for its exercise over society. Power was assumed to be territorialized in concept, even if the mechanics of its actual enforcement or outreach may have had more to do with the lines and networks of interaction. Another qualification might be that when we look at the early European states, they were hardly detached from the notion of a people sharing a common heritage. Despite the considerable movement and mixing of tribes and ethnic groups during the post-Roman period, many states saw themselves as the political embodiment of a *gens* or common people. It was an ideology which state rulers and the church actually fostered for their own ends despite the fact that it rarely had any factual basis in a strict sense (Reynolds 1983: 382).

In a discussion of how early states were formed, Renfrew has suggested that they came about through the interaction between and gradual merging of pre-existing units via military and political alliances. In this respect, they differ from empires whose formation usually involved the runaway growth of a core whose extension of hegemony was sustained by military force and whose relationship with peripheral societies was strongly asymmetrical. Renfrew suggests that the units which typically came together to form early states, or pre-state modules, usually involved a cluster of about ten such units whose individual size was around 40 km in diameter or approximately what could be reached in a day's journey from a centrally positioned focus of power. The aggregation of such modules into proto-states probably stemmed from the exchange systems that developed between them, with chiefs and tribal kings having a vested interest in creating flows of prestige goods and chiefly or royal women between their respective courts and lineages. When threatened externally, they would also have had a vested interest in coming together under a single military leader. The formation of states may have followed from the gradual fusion or coagulation of interests generated by such processes (Renfrew 1975: 12). However, we should also bear in mind Axelrod's 'TIT FOR TAT' concept. Using game theory, Axelrod demonstrated that where societies had only infrequent contact, or little expectation of re-contact, they tended to act in their own interests, or selfishly. However, where they had frequent or regular contact, as with Renfrew's clustering of early state modules, they tended to evolve principles of co-operation simply because they stood to lose as much as they gained from selfish or self-centred behaviour. Moreover, in the first instance, such co-operation need not be goal directed or managed in any purposeful or teleological way (Axelrod 1984).

To say that states were formed by the aggregation of pre-state modules is only part of the problem. The formation of complex hierarchical alliances was also a feature of chiefdoms and tribal kingships, but so too was the tendency for such alliances to collapse. What was different about early state systems is that they managed to endure, though some were more weakly coupled than others. I want to suggest that this new-found ability to persist was aided by the change which they gradually brought about in the conceptualization of political space. By shifting the burden of its definition from kinship and all its varied ties onto the notion of territory as the domain of a state ruler, the area over which he exercised absolute jurisdiction, territory was in a sense being defined outside of society. Its meaning was no longer confused with the social identity of the kin groups that occupied it, but was objectified as a domain of rule that could pass between rulers. In Europe, this redefinition of space was greatly aided by the development of feudalism. As a form of tenure, feudalism was a means by which state rulers could control the provinces. By granting out estates and territorialized blocks of jurisdiction to their followers or vassals in return for military service, state rulers created a system of governance whereby the provinces were ruled under their superiority. This superiority was underlined by the fact that when a lord or vassal died, his estate and jurisdictions passed back to the king. Though the more powerful territorial lords secured heritable estates and any jurisdictions held, the latter were still held from the king. The point which I am working towards here is that despite European feudalism's reliance on homage and vassalage, on the bonding of a man to his lord, the granting of land and jurisdiction in return for service fostered a more impersonal system of government. Through it, state rulers created a system of roles and functions that could be differentiated from those who actually held or performed them. Similarly, the space occupied by estates and their use rights or embraced by jurisdictions and their powers could be differentiated from those who actually enjoyed or exercised them.

It would be wrong, though, to say that in distancing government from kinship and in creating roles, functions and territories that were separate from the person who held them, feudalism created political structure. All pre-feudal systems had political structure too, but because it was largely embedded in the kinship structures of tribal groups and in their processes of success or failure, it tended to be a personal, volatile affair, indistinguishable from the processes that created it. What was different about early state systems is that by separating out roles, functions and spaces from the person and by making them a matter of reward for service, they created a notion of structure that could endure more easily across generations simply because their operation was now systemic or defined separately from the person who carried them out. It was as if part of the drama of human society had now became a play whose parts or roles could be played by different actors without it altering the nature of the play. It was no longer a case, as with pre-state systems, of actors playing themselves. In other words, the type of political structure created by early state systems had far more potential for inertia, simply because its structures and customary routines had become capable of being defined in a way that differentiated them out from those who controlled or exercised them.

Once the state acquired a structure and fixed routines of government that could be differentiated from its day-to-day practices, it was inevitable that this structure should be seen as a way in which the state was objectified. In short, whereas its day-to-day practices were about who exercised power at any one moment, its structure and routines of government were about its very existence. For this reason, we can hardly be surprised if the spatial structuring of the state, the mapping of its rights and jurisdictions into the landscape, should harden into fairly resistant forms.

In arguing for the early state as a different type of political structure, I am not suggesting it spawned wholly new patterns of political organization. In fact, when we explore early states, a prominent and striking feature is not how they redrew political geography but how they assembled existing parts into something different. Many European states incorporated 'territories' that had existed under pre-state systems, including Roman provinces, chiefdoms and tribal kingships, an incorporation that charged them with a new meaning and gave them a still greater resilience to change. However, for a number of European states, their emergence as stable political forms was not a simple, linear process. As with pre-state systems before them, what was put together in one generation might fall apart in another. Part of the difficulty was that despite the change in how political space was constituted, feudal states suffered from the fact that their integration rested on a balancing act. In order to control the provinces, state rulers had to make grants of lordship in return for military service. However, in the right circumstances, these provincial lords could use the power vested in them to assert a claim for more independence. The experience of France over the tenth–twelfth centuries is a good illustration of this point. Though the Capetian kings were claiming sovereignty over France by the tenth century, there was little force behind their claim. As the French countryside became divided into a dense network of local lordships, called bannum lordships because they exercised the ban or jurisdiction, the Crown's authority weakened to the point at which it threatened the existence of the would-be French state by the twelfth century.

Germany fared even worse. The original stem duchies of Saxony, Franconia, Swabia and Bavaria were welded together as part of Charlemagne's empire and later, after the Treaty of Verdun (AD 843), formed the basis for the East Frankish state. Further expansion to the east and, most notably, south into Italy led to its emergence as the Holy Roman Empire. As in France, though, the sub-division of lordship into a complex pattern of jurisdictions and the growing assertion of power by the lay landowners, bishops and towns that exercised these jurisdictions worked against the integration of a unified German state (Arnold 1992: 59–60). The Holy Roman Empire lasted down to the thirteenth century, by which time the German king or emperor had lost control over the Italian mainland and by 1232 had effectively conceded power back to the German duchies and princes, a loss of authority confirmed by the Golden Bull of 1356. At this point, Germany effectively was organized into about 150 separate states. Most were small except for the large eastern states like Brandenburg and East Pomerania, areas which had seen considerable change during the great colonization movements of the twelfth and thirteenth centuries. This extreme fragmentation of power actually increased over the next two or three centuries. By 1648, it was reportedly divided into 234 separate territorial units and fifty-one free cities. The remarkable fact is that when their unification was eventually achieved over the mid-nineteenth century, it merged political units that in one form or another had survived nearly 1,000 years of political process without being radically redrawn geographically. Indeed, even after unification, many survived on as units of administration within the newly merged German state.

In fact, when we review the long-term development of all the older state systems, there is a manifest degree of persistence about their basic structural form, particularly as regards how they were mapped into the landscape. Across western Europe, the provincial, regional and sub-regional districts around which early states became territorialized were based on areas that had a political identity or meaning under pre-state systems. Furthermore, as both church and state added a still finer scale of administration over the ninth–twelfth centuries with the creation of parishes and townships, or their equivalents, the local units of territorial order that were created developed into forms which have subsequently demonstrated an equal capacity for persistence. If we take a state like England which, along with Saxony, was one of the pioneer areas of European state formation, a case can be made out for seeing the territorialization of power and administration achieved during the early centuries of state formation as persisting in some form down until relatively recent times. This is not to deny the significance of subsequent change such as that brought about by sixteenth-century reforms in government. Elton described the changes of the sixteenth century as a 'revolution' rather than a 'transition' because they introduced 'bureaucratic management in place of the personal control of the king, and national management rather than management of the king's estate' (Elton 1953: 4). In the process, it greatly advanced the process of state integration, drawing all parts of the realm more firmly and directly under crown authority, albeit an authority now mediated through an increasingly specialist bureaucracy. To this end, the power of surviving liberties and lordships was abolished, from the numerous local manorial rights up to the powerful Palatinates of Durham, Lancaster and Chester (Elton 1960: 32). Where needed, as in Wales, new counties were introduced by the Acts of Union (1536 and 1542). In place of local lordships and the custom of local manorial courts, parishes were now adopted as the basis for a new fine-scaled judicial system, one that projected state power in a more uniform way. Yet as one study put it, 'England's precocious centralization around a comparatively strong crown' did not wipe the slate clean, but blended old and new (Corrigan and Sayer 1985: 17). Even Elton qualified his revolution in this way, concluding that 'a great deal had been simply preserved' (Elton 1953: 3) and that these medieval survivals continued 'a meaningless existence' down to the nineteenth century (ibid.: 424).

The next great 'revolution' in English administration consisted of the administrative reforms of the mid-nineteenth century and was precipitated by the rapid urbanization and changing distribution of population and, above all, by the burgeoning growth of industrial towns. However, as some commentators have noted, it was not a reform that anticipated or tracked such problems but one that involved a pragmatic and piecemeal response to them and then only after matters had reached crisis (Parris 1982: 281–2; Lubenow 1971: 180). Part of the problem was that despite its lead position as an industrial and urbanizing nation, the so-called workshop of the world effectively still had no 'system' or coherence of purpose in its local administration prior to the mid-nineteenth century, despite earlier 'revolutions' (Keith-Lucas 1980: 13). What existed represented a welter of old forms and instruments which, when need arose, had been charged with new functions and demands so that the overwhelming impression of its local government forms by the early nineteenth century was its generally makeshift and inertial character. Interwoven with these old forms and procedures, though, were a wide range of Statutory Authorities for Special Purposes devised on an *ad hoc* basis to solve specific problems (Smellie 1968: 18–19). Moreover, despite the push towards greater centralization provided by the Tudor revolution in government, it was still possible for some to see the lack of centralization in the nineteenth century as one of the weaknesses of a government now faced with the pressing need for a more co-ordinated approach to everyday problems and with the need for a more continuous exercise of government (MacDonagh 1977: 1).

These problems of anachronism were evident at all scales and in all areas of local government. Keith-Lucas talked about the 'extraordinary anomalies of the county system', with 'its wide variety of practices and customs' (Keith-Lucas 1980: 41). Matters were no different at the lowest scale of administration. Amongst the 15,600 parishes into which England and Wales were divided, some had no population and over fifty had less than ten inhabitants, whilst some had over 50,000, and each was burdened with responsibility, but hardly an integrated responsibility, for issues of poor relief, highways and policing (Lubenow 1971: 16–17). In the great industrial towns, where the problems of rapid growth were concentrated with dramatic effect, old units of administration greatly compounded the problem through their procrustean grip on the processes of government. Like the counties, said one writer, the borough system was 'in ruins' by 1830 (Smellie 1968: 22). Major cities like Sheffield and Birmingham possessed no unitary system or instrument of government until well into the nineteenth century. Indeed, Birmingham struggled to separate its interests from those of the surrounding county, still being treated as a township for rateable purposes until the 1840s! (Fraser 1982: 3). The anachronism was widely shared. Until the Municipal Corporation Act of 1835, the general verdict on town administration must be that it amounted to 'a mosaic of manorial, parochial, township and borough institutions, many of which were effete and most of which were hopelessly ill- equipped to deal with the problems' in hand (ibid.: 2).

Even when reform came, it was, typically for the times, piecemeal and drawn out. Change began with the Reform Act (1832) and the Poor Law Amendment Act (1834). The former began the process of shifting political power from the areas of 'past dominance' to the towns (Lawton and Pooley 1992: 19). The latter, meanwhile, had significance for town administration because of the way it created town-based authorities for poor relief. Further change quickly followed with the Municipal Reform Act of 1835. The 1835 Act did not affect all towns, or provide an immediate solution to all problems, but it did at least start the long drawn-out process whereby the structures and processes of general town management could catch up with some of the problems that were emerging, replacing patronage and oligarchies with more open, elected councils and executive officers. Through successive legislation, municipal boroughs not only acquired responsibilities but also new powers that enabled them to build a civic identity through the controls which they could exercise over housing, sanitation and sewage and the provision of services or facilities like schools, lighting, baths, wash-houses, libraries, museums, asylums and roads. Though earlier legislation had given them responsibilities for sanitation and health, the 1875 Public Health Act went further and turned them into public health authorities. Though anomalies and anachronisms still remained, the mid–late nineteenth century could at least claim that the structures and processes of administration were attempting to catch up with the demands now confronting them.

Modernization and societal change

The modernization of western society over the past two centuries is conventionally seen as being as much a socio-cultural as a political and economic transformation. Elaborating on the precise nature of this socio-cultural transformation is seen by some as a central problem for social theory, one that lies at the heart of the so-called Modernity Project.

If we look at the character of west European societies prior to modernization, what stands out is their fragmentation into relatively insular regional or even sub-regional communities that perceived themselves as different, as having a separate sense of community and identity, from those around them. Arguably, the basis for this regionalization was their *genre de vie*, the ways of life developed around the varied ecologies that are such a feature of so many parts of western Europe. Quite apart from the differences in livelihood and material culture that arose out of how different resource sets were exploited, there were other less visible differences arising out of the tendency for 'the self-enclosed community to communicate in terms whose meaning can only be identified in context' (Gellner 1983: 12). In other words, they were regions or sub-regions constituted as much in the mind as on the ground, as much in symbolic reference as in cultural practice. This regionally defined character of culture played a part in the structuring of the early state. When early European states were formed out of the aggregation of local chiefdoms and kingships, there was a tendency for their initial eruption to be based on what were broadly homogeneous core regions, regions that usually comprised fertile arable soils like the Ile de France and the Hungarian Plain (Pounds and Ball 1964: 24–40). By integrating like with like, state rulers were able to appeal to a ready-made ideological basis for integration and to introduce uniform institutional forms that effectively lowered their administration costs (Tilly 1975: 42). The homogeneous core of these incipient states helped to foster the belief that it was not simply a unit of socio-political control, but one inhabited by communities that shared the same origins and culture.

In time, as states expanded, this assumption of shared cultural character and common ethnic roots was projected out over areas whose actual cultural character could be quite distinct (Reynolds 1983: 382). Yet as an assumption, it was projected outwards at a particular level. Gellner's general model of how agro-literate polities were structured helps us understand how such an idea could be emplaced. His model draws attention to the lateral insulation of settlements and communities, each sustained within a particular ecology and enriched by a particular lifeworld. However, overlain across such settlements and communities was a unified capstone made up of continuous but horizontally differentiated layers of control, represented by military, bureaucratic, religious and commercial elites (Gellner 1983: 7). Like Redfield's notion of peasant societies as part societies or a combination of opposites, the so-called Great and Little Traditions, Gellner's model effectively divides the local community into a lower indigenous, introspective world, one rooted in the soil and its practical tasks, and an upper, intrusive but integrating world of prevailing power structures (Redfield 1960: 20). In western Europe, the power structures that bound otherwise disparate communities and ecologies over the medieval period were ordered around the institutions of military feudalism and the church. By their very character, though, feudal states could be fragile affairs, with the power devolved to territorial lords often fracturing the capstone of authority that was meant to bind the state together, leading to the state breaking up into relatively independent lordships as with twelfth-century France. In such circumstances, the unity provided by the church served the role of a more enduring capstone.

When we find European rulers trying to pull their states closer together over the sixteenth and seventeenth centuries, they were recovering their direct control over power that had previously been devolved to provincial and marcher elites. As a process of political change, though, it did not in itself remove the cultural differences that all extended states embraced. Indeed, through their stronger incorporation of peripheral areas and lesser states, as with England and the gradual steps towards a united kingdom, cultural differences were actually increased. When Corrigan and Sayer describe this process of early modern state consolidation in England as being as much a cultural as a political revolution (Corrigan and Sayer 1985: 2), they are really describing what it needed to be if it was to succeed. To this end, it was of necessity 'a totalizing project, representing people as part of a particular community' (ibid.: 4). The 'totalizing project' was not something that could be legislated for in any simple way, for quite apart from any practical changes in how people were governed, it required that disparate groups within society shared the same sense of identity and belief. The centralization of power in the hands of state rulers, and the focus now given to the processes of law and matters of administration, certainly served the cause of national integration, though they hardly made it a *fait accompli*.

Of course, other ongoing processes at this point were just as capable of smothering cultural difference. The progressive improvement of communications was a powerful facilitator of cultural integration if only because such improvements invariably emphasized the paramountcy of the centre, addressing 'its symbols and problems' (Eisenstadt 1963: 99–100). The growth of trade, both internal and external, was another important factor. Yet though some rightly see trade as generating a form of cultural as well as economic integration through its capacity for spreading the values of the marketplace (Hill 1969: 27), others see it as capable of creating a broad sectoral divide between areas of old and new wealth, a divide that was also profoundly cultural. Fox's division of early modern French society and space into a land-based culture resting on the residual power of feudal magnates and a maritime- or Atlantic-based culture centred around the wealth of a newly expanded commercial sector certainly has cultural undertones (Fox 1971: 173). In fact, as Fox himself was acutely aware, there were other parts of Europe where the cultural differences between areas dominated by feudal lordship and those dominated by commerce were much more deep-rooted (ibid.: 178). He saw Europe itself as partitioned by a broad belt of commercially based city-states and trading towns that stretched from northern Italy across the Alps, with a branch extending into the Low Countries and another into Germany, incorporating the trading towns of both the Rhenish and Hanseatic Leagues. Fox saw this belt of independent and semi-independent towns – or Rokkan's city-state band (Rokkan 1975: 576) – as comprising societies that were different socially and culturally from the land-based feudal societies that lay on either side. Largely because of these differences, the expansionary drift of land-based empires signally failed to absorb such towns, leaving them as a different sort of cultural as well as economic space 'between, rather than within, the rising monarchies' (Fox 1971: 33; Rokkan 1975: 576).

By the eighteenth century, the part played by the Enlightenment in changing society's approach to knowledge was of fundamental importance to socio-cultural change, as were the changing means by which new knowledge spread both geographically and socially (Thrift 1996: 96–124). What had been treated as unproblematic, as explained adequately by the teaching of the church, now became open to debate. The natural and, later, the social world, became objectified, opened up to new forms of knowledge creation and validation that seemed, at first, norm free. Though initially part of a very limited cultural sphere, the ideas of the Enlightenment paved the way for a more rationalized lifeworld, one that was used to build a rival world not just to that offered by the traditional teaching of the church but also to that rooted in the vernacular world of tradition and mythology. Though the age of reason was first manifest in ideas about the physical world, the nature of society itself was also opened up to a more rationalized form of enquiry by the late eighteenth century (Bowen 1981: chapter 6).

Such developments might have been confined in their effect had it not also been for the changes in education which swept through European society over the eighteenth and nineteenth centuries. Perhaps more than any other single change, the introduction of standardized educational systems, particularly over the nineteenth century, served to produce an ideology of belief about the nation-state and its common culture. Rival languages, and the alternative ideologies that were invariably embedded in them, were now actively discouraged as states strove for a new degree of cultural homogeneity. Needless to say, this was especially the case in states like Britain and France which found themselves with a number of alternative languages inside their borders and which perceived them as a threat to their integrity as nation-states (see, for example, Withers 1988: 110–66). However, of all the changes wrought by developments in the state provision of education, the one that mattered for the erosion of cultural differences and for the implantation of a single reading of the state and its origins was the spread of literacy. Once its message could be unlocked through literacy, it was the power of the written word which, more than any other single factor, was able to 'destabilize' traditional societies (Todd 1987: 144). Literacy brought the majority of adults, especially male adults, 'within reach of ideological texts and concepts that in a traditional society belong to a tiny elite' (ibid.: 136). More than any other change, its spread opened up the possibility of a homogenized national culture by giving access to the more standardized presentation of knowledge afforded by the printed word (Thrift 1996: esp. 113–14), though, by way of a qualification, it could also be argued that high levels of literacy have actually encouraged a more interpretative reading of texts in recent times, whilst the growing specialization of knowledge has opened up new and profound differences in the type of knowledge available to individuals.

By the mid-nineteenth century, other factors were at work. The high levels of migration that accompanied rapid rates of urbanization and industrialization redrew the internal geography of west European states, bringing together large and diverse communities of people and creating wholly new conditions for the subsequent acculturation of their offspring. Gellner has drawn attention to how this process of large-scale urbanization, with its high levels of mobility, transforms the orderly, stratified society of the pre-industrial world into a world in which social entropy or disorder is increased, 'a shift from pattern to systematic randomness' (Gellner 1983: 63). Reduced to essentials Gellner's point is directed at society's loss of internal structure during the process of modernization as vernacular cultures are replaced by the anomie of the city. Urban industrial society forms a more *ad hoc*, fluid system, one lacking the traditional sub-structures through which individuals mediate with the state at large.

His point owes much to the thinking of Habermas and others on how the processes of state integration coupled with large-scale industrialization and urbanization have not re-aligned human cultures so much as dislocated them. Habermas addresses the problem at its most fundamental level, or how changes in the nature of society altered the relationship between the internal perspective of the lifeworld, on which communicative action is based, and the systemic forms and roles through which society at large is integrated. For the purposes of analysis, he sees their relationship as changing via two broad adjustments. At the outset, with tribal society, society was organized through kinship in a way that did not disrupt the structures of the lifeworld, their inner and outer world sitting comfortably one with another simply because one was seen as a more distant extension of the other (Habermas 1987: 154 and 188). With the shift from tribal societies to state systems, though, new forms of integration were institutionalized that now lay beyond the lifeworld and which, for that reason, raised new problems of validation or legitimation. The structural constitution of polities in the form of the territorialized state can be seen as the cornerstone for these new forms of integration, 'an objectified context of life' to use one of Habermas's phrases (ibid.: 173), that developed outside of the norms and values of the lifeworld. With modernization, and the emergence of still greater levels of functional differentiation and system complexity, the problems of finding 'an independent morality effective for integration' or 'new normative orientations capable of securing an organic form of solidarity' (ibid.: 116) have grown apace, creating a legitimation crisis for modern society (see also, ibid.: 172 and 188).

Without denying the importance of such arguments, though, they cannot be taken as the last word on how modernization has impacted on culture. Despite the massive social, economic and political changes that have driven modernization, the remarkable fact is how much of traditional culture has endured. In particular, there is a case for suggesting that whilst the shift towards a more rationalized lifeworld has contributed to the cultural transformation of society, there are still normative structures within the lifeworld that have changed little.

In fact, a systematic analysis of human culture's inertial properties has been provided by Todd in his cultural interpretation of political ideologies. Reduced to essentials, Todd's thesis is that 'the ideological system is everywhere the intellectual embodiment of family structure' (Todd 1985: 17). Drawing on ideas first advanced by Frederic Le Play, he develops differences in family structure around two axes of variation. For Le Play, relations between father and son were thought to shape a society's concept of liberty or the opposite, whilst those between brothers were seen as shaping its concept of equality or the opposite (ibid.: 6). Different societies can be aligned along these two axes (liberty/authority and equality/inequality). When considered in a European cultural context, such a scheme generates four broad types of family structure: the absolute nuclear family (liberty + inequality) and its opposite, the community family (equality + authority); the egalitarian nuclear family (liberty + equality) and its opposite, the authoritarian family (inequality + authority) (ibid.: 10). Despite the cultural uniformities that have prevailed in many parts of Europe (i.e. Christianity, feudalism), these different types of family structure produce a fairly complex geography of variation, with adjacent countries having different family structures. Thus, Todd sees England as characterized by the absolute nuclear family, northern France by the egalitarian nuclear and Germany by the authoritarian family (ibid.: 16). Perhaps influenced by Braudel, he roots such differences not in their overall cultural experience but in adaptations that were made when these societies first settled the land in a permanent fashion back in prehistory. Having rooted their origins in the Neolithic/Bronze age, Todd proceeded to bring their significance forward into modern times by arguing persuasively that their differences in family structure underpin their differences in political ideology. Nor is it an argument confined to the arena of Europe. He also applies it to global differences. If accepted, it would mean that normative values at the most fundamental level of the lifeworld have not only influenced the system integration of societies but, contrary to Habermas, have continued to do so despite all the changes wrought by modernization. The sheer depth of inertia which Todd attaches to family structures may be questionable, but his case for the cultural permeability of modernization is well made, as is the fact that its impact varies geographically. It does, of course, have echoes in Berry's suggestion that even in a sphere which has supposedly been transformed through wholly new values – the economic – and which, more than any other, articulates the global at the expense of the vernacular, nevertheless, we need a 'culture-specific economic geography', one which recognizes that economic differences have 'a cultural foundation' (Berry 1989: 17–18).

Modern capitalism and the problem of regional change

The rise of modern capitalism from the late eighteenth century onwards, with its potent fusion of industrial and market capitalism, was hugely significant for the regional experience of societal change. We can characterize its impact in three ways. First, it led to the mechanization of production systems within the framework of the factory system and with it, the introduction of new work disciplines. These initiated far-reaching changes in both the processes of production and in the way industry was organized, changes that were to prove an ongoing if cyclical process (Schumpeter 1971: 37–9; Gregory 1990b: esp. 356–72). New and more productive technologies or techniques of manufacture were introduced, making production processes progressively more automated, larger in scale and cost effective. Additionally, new and more productive ways of organizing production were also sought that made it increasingly more integrated and efficient. The sharp upward shift in the size of British industrial units, especially lead-sector industries like cotton production and iron-working, during the 1830s and 40s, was one such adjustment.

Chandler's work on the history of American industry has documented the organizational changes that were so critical in raising its output and productivity over the late nineteenth and early twentieth centuries, some of which were not classifiable as 'innovations' in a Schumpeterian sense (ibid.: 136). The emergence of large multi-function and multi-product firms especially, involved new problems of organization (ibid.: 127). Commenting on the increasing amount of growth through merger, Chandler drew a significant contrast with Britain. Mergers in the latter tended to form federated combinations. In America, meanwhile, they led to elaborate 'managerial hierarchies to coordinate, monitor and allocate resources to the constituent firms' (Chandler 1988: 421). Such organizational changes provide the context in which large industrial corporations in America began to adopt parts standardization and assembly-line techniques by the 1920s and 30s. With the great expansion of market potential after the Second World War, this system of organizing production in large, upwardly integrated and increasingly international companies that were mass producing standardized goods using assembly-line techniques, or Fordism, was expanded rapidly, not just in America but also in parts of north-west Europe (Storper and Scott 1989b: 25). Their sheer scale of operation, and their need for a wide range of inputs and supporting services, invariably meant that such firms formed the centre for a whole complex of interlocked activity (ibid.: 25). By the 1970s though, as crises began to threaten the continuing success of Fordism, reactionary changes began to occur, with some industrial corporations adopting a 'hollow' form of organization, a more flexible form of production, with more fragmented production units and with equipment and labour used in more flexible ways so as to enable production to be expanded or contracted quickly and to be shifted easily between products (Swyngedouw 1989: 37; Storper and Scott 1989b: 24).

A second feature of modern capitalism is its association with free or price-fixing markets. It was Polanyi who first made the point that the beginnings of the factory system helped to force open regulated markets. The rise of production systems in areas and sectors that were not covered by medieval regulation did much to undermine the significance of regulation. For Polanyi, it was the scale at which industry now operated that acted as the main solvent of what remained of regulation. As the scale of its operations increased, so did the risks attached to investment. It needed regular supplies of land, labour, raw material and capital on an increasingly larger scale. To grow unchecked, it needed to force such factors of production into the market place and to be able to bid for them. Further, as the formative influence on the expansion of urbanization, it had a vested interest in having an adequate and cheap supply of food for its labour force. In consequence, it also had an interest in supporting a free market for agricultural produce. Finally, it had a vested interest in opening up markets that were closed to its products or protected by tariff walls. It was not surprising that at the end of the Napoleonic Wars in 1815, after industry on the continent had been cut off from the technologies being pioneered in Britain, industrialists in the latter favoured freer trade and the chance of competing in European markets.

The third way in which we can characterize modern or industrial capitalism is through the increasing and more intensive use of capital. Its growing appetite for investment, though, created a problem, one that struck at the very heart of capitalism as an economic system. This problem can best be appreciated in terms of how it evolved. Medieval and early modern systems of industrial production largely involved domestic workshops and simple, hand-driven equipment. The levels of fixed capital involved were modest. Even where merchants had built up elaborate, extended systems of putting out, the bulk of the capital involved comprised circulating capital or that embodied in the raw and semi-finished materials which, at any one moment, were to be found moving through the system. Though some commercial enterprises in the metalliferous industries built up sizeable aggregate investments in scattered mines, furnaces and forges, the units involved were relatively small as was the capital locked into any one site. With the Industrial Revolution, and growth of the factory system, the level of fixed capital began to increase. The change was not immediately dramatic. The early factories were often adapted from other uses rather than custom built and the machinery introduced was generally wood- rather than metal-based and engineered. By the 1830s and 40s, with the construction of more purpose-built integrated factories, key industries like cotton and woollen-cloth manufacture saw the level of fixed capital jump to between 30 and 40 per cent of the total capital involved. By the late nineteenth century, heavy technology-based industries like chemicals had experienced a further surge of capital deepening. By the early twentieth century, and the construction of machine-based assembly-line systems, this capital deepening spread to a range of new industries. Alongside these progressive increases in the proportion of industrial capital that was fixed, there also occurred massive investments in the infrastructure of the industrializing economies. Railways, for instance, consumed a significant proportion of the gross national product in countries like Britain, Belgium and America over the middle decade of the nineteenth century. Equally large amounts of fixed capital were sunk into urbanization and into public utilities like roads, water, gas and electricity supplies. The sum effect was that by *c*.1900, the main industrialized countries had landscapes overwritten with high levels of accumulated fixed investment.

In the long term, this build up of fixed capital investment, and its increasing proportion of total capital investment, was not sustained. However, the levels which it did reach created a different kind of geography, one that posed critical problems for the circuits of capital within which landscape was now heavily implicated. Again, we can best make the point by drawing out the comparison with earlier forms of industry. So long as industry involved a high proportion of circulating rather than fixed capital, merchants could exercise control over how capital could be invested at the end of each cycle of investment since their capital was returned to them with interest at the end of each unit of account. If new opportunities for investment arose, the merchant could easily redirect capital to them. However, with the sharp scalar growth of fixed capital investment from the late eighteenth century onwards, this flexibility was lost. At first, its loss was not perceived as an acute problem, for capital was provided through the business partnerships that also ran them. Fresh capital needs were met by the reinvestment of profits for, in the circumstances, such plants were seen by business partnerships as an investment of entrepreneurship as well as their capital. They embodied the know-how and business acumen of the partners as well as their capital. By the mid-nineteenth century, though, conditions had changed. The scale of capital required had risen beyond the capacity of small business partnerships. New institutionalized forms of provision were provided through joint-stock companies and merchant banking, though initially these were more successfully exploited as sources of investment in countries like Belgium and Germany than in Britain. However, the penetration of investment capital into industry brought capital's ongoing need for sustaining sufficient returns by way of interest into direct conflict with the loss of flexibility caused by its growing embeddedness in fixed forms. The conflict was accentuated by the fact that as industrial output became more and more tied to specific sets of technology and to the success of particular products, its profitability became tied to cycles of investment in new technology, new products and the advantage which an investment more attuned to market conditions could bring. To put the point more simply, new investments, being best-practice, tended to enjoy above average rates of return. Slowly though, rates of return would average out.

For the now autonomous or independent capital market, the solution to this problem of declining returns was for such investments to be amortized, written off at a fixed rate per annum so that, after a period, the original investment was recovered as well as interest for the loan. It must be stressed that this apparently simple solution to the problem of investment in plant that was both technologically and geographically fixed was not arrived at in any simple or linear way. Like the academic debate of how the relationship between working and fixed capital should be computed, various solutions were possible (e.g. Hicks 1973: 8).

Once solutions had been devised, capital acquired a contrived flexibility as a higher circuit of investment capital came into being with the ability to follow the market and to seek out new technologies, resources and spaces, investing and disinvesting at will. In the process, though, regions or sectors of existing investment, with all their accumulated but tied investment in particular manufacturing technologies, transport systems, housing and labour skills, were abandoned to slow decline. As Harvey and others have pointed out, there is an obvious and profound significance in this mobility of capital for the geography of the built environment. Communities, and the built environment on which they depend both for living and working, are created through particular cycles of investment. For a time, they flourish as the capital invested in them reaps the benefit of better than average returns from better than average technologies or lower costs. In time, though, as new investment cycles redraw production functions through new areas of investment, old areas decline. Few established centres of investment are able to overcome relative decline by making the kind of radical change that comes more easily to wholly new areas and sectors of investment (Harvey 1985: esp. 1–61).

A case against such an argument can be put together both conceptually and empirically. Thus, Myrdal's theory of cumulative causation assumed that once regions of growth gained advantage within a system of regions, then they tended to reinforce not lose that advantage as the benefits gained from their success were used to compound their advantages, to load the dice in their own favour (Myrdal 1957: 11–22). A study like Pred's work on the American city-system can be seen as offering empirical support for such an argument. Looking at the rank-order of cities over time, he demonstrates how those with lead advantages during the early phases of the American Industrial Revolution were still in dominant positions by the 1950s and 60s. The conclusion that he reached was that 'long-term rank stability of large metropolitan complexes . . . can be most plausibly explained by the tendency of early established major channels of interdependence . . . to be self-reinforcing' (Pred 1977: 36–7; see also Arthur 1988: 85–97).

Pred's analyses took a long-term perspective, one probing back deep into the nineteenth century. They were also focused on the flow of information rather than on patterns of change in the economy at large, though obviously, the two are not unrelated. However, the general implications of his argument here have to be seen beside a range of other work on recent trends in America which have emphasized the inter-regional shifts that are now taking place. At the heart of this work is the shift seen to be taking place from the 'rust- or snow- belt' to the 'sun-belt'. The traditional industrial core regions of New England, the Mid Atlantic and the East–North–Central census regions have lost ground to the more peripheral regions, particularly the Southeast and Southwest regions, or the so-called sun belt (Rees 1979: 47). Rees portrayed these regional shifts as 'a series of core–periphery realignments, a manifestation of spatial decentralization tendencies' that are playing themselves out at different scales of the system from the regional to the international (ibid.: 47). Rees's study compares closely in its conclusions with that by Varaiya and Wiseman on the 'age of cities'. What they call the 'vintage-profile of a city's productive capacity', its balance between the retirement of obsolete equipment and investment in new, can be used to date cities (Varaiya and Wiseman 1978: 127). Applying this to American cities since 1947, they seek to explain the broad shifts in manufacturing through the growing age, or vintage profile, of capacity in cities situated in the older manufacturing areas. The older their average stock, the more they respond weakly to booms and cope badly with slumps (ibid.: 128). In a discussion of a still more recent phase of change, or the switch from Fordism to forms of flexible accumulation, Scott proposed that the initiative had been seized by areas which, over the 1950s and 60s, had been regarded as 'marginal areas' and as 'overshadowed by the main centres of production' (Scott 1988: 3). Such long-term shifts in capacity and performance have obvious implications for interpretations of change rooted in some form of cumulative causation theory. In fact, having the broad shifts of employment capacity from the snow belt to the sun belt and from large urban agglomerations to smaller ones in mind, Casseti thought it directly contradicted any ideas based on the principle of cumulative causation (Casseti 1981: 572). Just as mindful of what these shifts might mean for the wider problem of change, Gertler has also used them to question the validity of cumulative causation, arguing that it 'does not appear to be a viable construct in the current endeavor' (Gertler 1984: 172).

These inter-regional shifts can be seen elsewhere in mature or maturing economies. Reviewing the relationship between technological innovations, investment phases and urban growth in Britain over the nineteenth century, Robson talked about the 'rise and fall of particular places' being 'a function of the individual patterns of investment, disinvestment and reinvestment in given locations' (Robson 1981: 127). The shifting pattern of investment in particular industries over the late eighteenth and nineteenth centuries had already been apparent from earlier work on key industries like iron production (Roepke 1956). What Robson's work did was to draw out its wider implications. Linking changes in the geographical pattern of investment to the history of individual regions has also been a central thread of argument in Massey's work. Concentrating on changes that have taken place in the organization of the British economy over the twentieth century, her work deepens our grasp of this problem further. She notes that particular places are locked into wider spatial structures and wider spatial divisions of labour, but argues against seeing this relationship as the product of a simple or single phase of investment. When we look at the economy of particular regions, they appear 'as the historical product of the combination of layers of activity', layers which 'represent in turn the succession of roles the local economy has played within wider national and international spatial structures' (Massey 1984: 118). With each layer of investment might have come 'different organizations of dominance and subordination' (ibid.: 122). However, she sees these changing roles as not about whether regions or places are growing or declining, as if passing through a life-cycle of change, but about how they related to the particular needs of capital during each phase of investment. Yet whilst she argues against reducing the character of regions to the metaphor of a life-cycle because of their composite character, she accepts that each layer of their history may involve spatial structures that are at different phases of establishment, growth and decay (ibid.: 122). Such an interpretation clearly complicates our grasp of the experience or history that is built into particular regions and their configurations of capital, as it was intended to do. Ultimately, what Massey is anxious to stress is the uniqueness of these configurations and the inadequacy of employing general deterministic models to account for them, a leitmotif of much recent writing (cf. Dear and Wolch 1989: 11).

Faced with this temporariness of patterns of investment and employment, and the fact that change adjusts geographically as well as temporally, some writers have invoked the operation of purely economic factors. They base their explanations around the concept of the product cycle and its ability to generate regional cycles (e.g. Rees 1979: 51). Others have made use of changes in marginal productivity, with both relative and absolute levels of productivity and even output declining as the capital stock of a mature region, its inherited capital stock, becomes more obsolete (e.g. Varaiya and Wiseman 1978: 128; Rees 1979: 51; Casseti 1981: 574–5; Gertler 1984: 151). However, there is another dimension to the problem, one that provides an explanatory context for changes in marginal productivity. New investments in fixed capital stock can replace old or prior forms but are not substitutable *in situ* except without considerable extra costs, costs that increase in older, denser centres of industrial investment simply because of the greater embeddedness of capital in any mature system of output. In other words, upgrading production functions through new investment has to cope with the shape and organization of prior forms in a way that wholly new spaces of investment do not (cf. Massey 1984: 122). It is through the interaction of old and new, and the constraining power of the one over the other, that mature regions and cities appear to experience eventual if gradual decreases in marginal productivity as the rising costs of change deflect new investments and as the amount of change actually achieved becomes more marginal (Casseti 1981: 574; Gertler 1984: 151).

I have tried in this and the previous chapter to review different types of change and to do so at different scales. Throughout, I have deliberately given emphasis not to how easy change is but to how problematic it may be. Whilst we can be impressed by how some empires, world systems and regions maintained their position as centres of power, wealth and innovation for lengthy periods, the general conclusion must be that advantage is never sustained. Whether seen in terms of the costs attached to administration, military control or economic systems, there is a range of evidence to suggest that sustained growth ultimately leads to diminishing returns on continuing investment in existing forms. Furthermore, even when attempts are made to change or revolutionize existing forms, compromise abounds, with existing structures, technologies and routines, wholly or partially, being adapted for new purposes. In these circumstances, it is difficult to argue the case for society as a continuously adaptive system. Society may reflexively monitor the conditions of its own existence, but it does not follow that it invariably responds freely to fresh needs and opportunities. No less interesting, though, and not unconnected with this point is the extent to which radical or qualitative change can be shown to have frequently erupted around the edge of established systems at all scales, from empires and world systems down to recent industrial complexes. In effect, change appears to have a geographical dimension, one that redraws advantage and disadvantage between core and periphery in a periodic fashion. Clearly, we need to consider how the stagnation or decline of core areas and growth of peripheral areas may be connected, not least because it offers the prospect of an intrinsically geographical understanding of change. If we are to develop such an understanding further, though, we need to clarify how the geography of society may be implicated.

I want to suggest that the way in which society organizes itself in space forms a powerful source of inertia for societal systems, one that ultimately retards or deflects change. To establish a rounded or comprehensive grasp of change, therefore, we need to begin with a clearer, more systematic understanding of how the geography of societal systems acquires such inertial properties and how these inertial properties are not just mapped into their geography but are their geography. Altogether, I want to examine three different types of inertia over the next three chapters: cultural or symbolic, organizational and that created by the built environment.

5

Sources of inertia: the cultural construction of landscape

    ... thought reveals that man is not contemporaneous with what makes him be
    – or with that upon the basis of which he is   (Foucault 1974: 334)

As a characteristic of societal order, inertia is a property of its structural character. In a sense, it helps define structure, for those aspects of societal order
that change slowly, which resist easy change, are the very characteristics
around which its processes and everyday practices are necessarily structured.
To bring the problem of inertia into sharper focus, therefore, is to recover a
concern for the problem of structure as something that can be defined and
treated separately from societal processes and practices (cf. Lloyd 1993: 4).
Any analysis of this inertia must begin with the so-called deep structures of
society or, to use Bourdieu's term, its cultural and symbolic capital (Bourdieu
1992: 97–108).

    Most definitions of culture, including those which stress the materialist
aspects of human behaviour (i.e. *genre de vie*, adaptation, skills) and those
which stress the human capacity to symbolize or attach meaning to what is
around them, share one overriding characteristic: they see culture as embracing those traits, values and beliefs that endure or are carried across generations
and which, for that reason, are accepted by a society as given. Clearly, such a
definition, by stressing its inter-generational aspects, comes close to equating
culture with what is inertial about society. Yet though culture can be seen as
time-laden when viewed from the outside because of the way its forms persist
through time, when we invert the perspective and consider it from within, it
actually seems to collapse time into a shallow or limited concept of meaning
by denying any corrosion or loss of substance through time.

    Early cultures especially, have no real sense of chronology but treat all past
time as variants within a single phase of experience. For obvious reasons, they
have a sense of cyclical or seasonal time, but this is treated as time that is continually being recovered because its forms and routines are continually
repeated or recreated. It is not allowed to become history, to admit change, by

being allowed to offer a different or irreversible experience. When we consider early culture's conceptual handling of time alongside its handling of space, we find revealing contrasts. Early culture necessarily held an elaborate understanding of its spatial or environmental context simply because it was of the essence to its problems of livelihood and survival. Society could not afford to be disinterested. The fact that it was not explains why so many of the traits and beliefs of early culture relate to the practical problems of interacting with its material world. However, it did not map or construe this world in a pure Euclidean sense. This does not mean early cultures were incapable of producing maps (Leach 1976: 51; Lyle 1993: 64–8), only that the maps which they constructed were deeply topological, transformed by a rich overlay of meanings and associations. These meanings and associations were derived in large measure from the way they devised interpretations using classificatory schemes and cosmologies that brought the physical and social dimensions of their world together in a conceptual or affective sense. For this reason, ongoingness or persistence – and with it, the suppression of time and change – was seen as best achieved by the continual recreation of their material world through concept and ritual. The sum effect was that much of the inertia which permeated early culture was really about preserving the world as they conceived it, the ideological maintenance of their cultural landscape with all its overlays of symbolic meaning.

Denying history: the inertia of early culture

At first sight, the way early culture appears to make an adaptation to the problems of living in its environment may seem the most logical approach to establishing how early cultures became inertial. Defined in relation to human culture, adaptation represents a commitment of organization and skill to the problems of energy capture within a particular environment or ecology. In most evolutionary models, it is assumed to be a relatively fixed or inflexible commitment, because it is a response constrained either by the limited possibilities of its environment, the limited possibilities offered by the knowledge and skills of the group exploiting it, or by a combination of both. Being a fixed commitment, such adaptations are seen to bear on the problem of change in a passive way, with only the fittest solutions surviving the stresses and strains posed by environmental variability through time. In some circumstances, though, we can add an element of competition, as groups clinging to weak solutions fail and those holding more successful solutions thrive and expand.

In the literature, such a reading of adaptation is mostly developed in relation to the survival strategies employed by the earliest cultures. As an argument, though, it may overstate its case. Some would argue that the earliest or most basic forms of human adaptation have sufficient flexibility to cope both with geographical and temporal variability. Far from being 'homogeneous, stable, or conservative', the adaptive strategies of less complex societies were capable of being 'dynamic' simply because most evolved adaptations dealing with different environments (Denevan 1983: 401). Moreover, and a critical basis for this responsiveness, cultures dependent on basic strategies usually have a prior store of possible solutions that can be drawn on to cope with change (ibid.: 402). In fact, contrary to what might be thought, adaptation is more likely to be a source of inertia with more advanced societies. It was only when cultures began to invest or accumulate large amounts of energy in specific strategies, particularly strategies which involved effort spread over a number of generations such as with the early hydraulic civilizations, that we can begin to speak of the cultural adaptation to fixed forms as a likely source of inertia.

Another qualification to early cultural adaptation as a direct source of inertia concerns the problem of how it is defined. It is misleading to see it as involving the optimal or the best possible adaptation under the circumstances. In reality, many cultural adaptations are better described as being sufficient under the circumstances. As Hallpike has remarked, a great deal of early cultural 'adaptation' may have involved the survival of the mediocre rather than the fittest, or traits that had only limited functional value in their environmental context (Hallpike 1986: 113). Their persistence, therefore, could hardly be described as the outcome of a particularly successful adaptive strategy.

For early cultures, inertia was more likely to originate in the mind and in how they perceived or symbolized landscapes. Structuralist thinking especially, can guide us here. For the linguist Chomsky, basic attributes or rules of language are not reflexively developed, or culturally constituted through a trial and error form of practice, but are structure-dependent (Chomsky 1976: 34). Given the role of language and its nuances in the constitution of the life-world and its effect on the ability of individuals to reach a mutual understanding, we cannot ignore the possibility that there are aspects of the problem which may be 'structure-dependent' in a way that places them beyond easy revision. Indeed, Chomsky himself argued that through language, his 'innateness-hypothesis' as he calls it could include 'principles that bear on the place and role of people in a social world, the nature and conditions of work, the structure of human action, will and choice, and so on' (ibid.: 35). What makes Chomsky's ideas doubly relevant to us is his further suggestion that this 'innateness-hypothesis' has already been validated in regard to other aspects of human cognition, including the human construction of visual space. Reduced to essentials, the world of Chomsky is a world in which there are biological constraints both to how we perceive the world and how we articulate our thoughts about it. Within the span of human history, such constraints can be taken as relative constants.

As a basis for inertia, structuralism can be developed in other ways. The anthropologist Lévi-Strauss has posited an array of ideas on what he saw as the structural aspects behind early culture's capacity to symbolize its world. Like Chomsky, he saw the rules involved as having an 'innateness' about them, but he does not see them as necessarily being genetically programmed. Some are culturally constituted. Yet though culturally constituted, he does not see them as reflexively derived, nor as rules that vary between cultures. They are deeply based structures of thought, acted upon but not consciously rationalized in any way. They are recoverable by systematic research but are not articulated in their entirety by any single myth or ritual. Furthermore, though the manner or cultural context in which they are articulated may vary between cultures, the basic nature of the rules does not. It is in this sense that they can be regarded as manifestations of human culture's deep structural form, something that takes us back to the very foundations of culture itself.

Altogether, Lévi-Strauss developed his thinking in relation to three types of rule. First, there is the basic problem of how primitive society orders its world through classificatory schemes. Like all classificatory schemes, primitive or early classification was designed to differentiate. At root, he saw the schemes employed as binary, classifying all meaningful elements of the world (clans and moieties, plants, animals, physical features, colours, etc.) into a series of simple oppositions, such as earth–sky, abundance–scarcity, red–black and so on, each of which can be seen as homologous to each other. Around this basic binary code is constructed a scheme capable of infinite extension, one that can internally differentiate the world *in toto* but in a way that establishes practical and theoretical relations between the parts. More important, Lévi-Strauss sees the totemic nature of such schemes, with classificatory equivalences between social groups and animals or other features of the physical world, as misleading if it is read as an attempt to classify equivalences of content or form. It is not a classification of humans, animals or plants *per se* but a classification of the world through the differences perceived and understood between them and their various characteristics. They provide a metaphorical basis by which the world can be ordered and its parts related. As Lévi-Strauss himself demonstrated in a variety of studies, including work on the symbolic layout of settlements, the use of such classificatory schemes means that the cultural construction of the world, the way they are used to order and interpret the cultural landscape, was fundamentally shaped by structures of mind that, by their very nature, were deeply embedded, inertial in character and widespread in their distribution (Lévi-Strauss 1966: 135–60; 1968; 1969: 69–83; Griaule and Dieterlen 1954: 83–110; Dodgshon 1985: 61–83).

A second dimension of culture which Lévi-Strauss subjected to structural analysis, one that also has significance for how society organized itself in space, is that of kinship and marriage alliance. Again, he stressed how the variety of solutions adopted by different cultures and societies may, when analysed structurally, reveal the operation of a few basic rules. Drawing on ideas from communication theory, he depicted early cultures as generating their kinship systems under conditions of constrained rather than free choice, hence the manifestation of patterning rather than randomness.

A third broad theme running through Lévi-Strauss's work is his interest in the interpretation of myth. Like other aspects of human culture, he saw myths as structured around deeply held anxieties or concepts, such as the distinction between nature and culture or the problem of incest. Such concerns can be articulated via a whole series of myths and ritual practices. An imaginative illustration of such anxieties and their explanatory power has been made by Hodder. His study draws attention to how the development of farming needs to be seen within a framework of durable symbolic structures that existed before farming and provided a 'formative' context for its development (Hodder 1990: 280–2). His core thesis rests on the deep-rooted human concern with the basic distinction between culture and nature. For Hodder, the house or *domus* became symbolically contrasted with nature, a space and world opposed to the wildness of the latter. In time, with the advent of farming, the symbolism that surrounded the *domus* became the central metaphor in the taming of the wild. The Greek word *agros* meaning field is associated with the adjectival form *agrios* meaning wild, so that we can see the word agriculture as meaning the taming or culturing of the wild (ibid.: 86). The symbolism surrounding the *domus–agrios* opposition is seen by Hodder as not just a context within which farming evolved but as one extended and used as a metaphor for social domestication and for different forms of dominance and subordination within a network of social relations. Though concerned with a fundamental change and all the contingent social and political changes that followed, Hodder offers the concluding thought that his study is about 'long-term continuities in Europe' (ibid.: 307).

In fact, he is tapping into a problem that extends beyond Europe. As Wilson put it, 'cosmocising houses is remarkably widespread among traditional peoples', a practice that derives from the concept of the temple as the centre, the *axis mundae*, of the world (Wilson 1992: 111; see also, Wheatley 1969). Bourdieu's study of the Kabyle in Algeria provides one of a number of such studies that might be compared with Hodder's symbolization of the house as part of a *domus–agrios* opposition. The Kabyle house was 'organized in accordance with a set of homologous oppositions – fire:water/ cooked:raw/ high:low/ light:dark/ day:night/ male:female/ *nif:hurma*/ fertilizing:able to be fertilized/ culture:nature' (Bourdieu 1979: 141). Such oppositions formed a *principium divisionis* by which the house or farmstead and the universe at large could be drawn within the same system of order (see also, Dodgshon 1988: 69–82). Some would see this symbolization of the house as a microcosm of the universe and its oppositions as a wholly cultural construction, one consciously and deliberately arrived at by early culture. Wilson, for example, explains it via Eliade's point that 'one cannot settle in the world without assuming the responsibility to create' (Wilson 1992: 111). Yet for the structuralist, there are aspects, such as the use of binary oppositions and homologues, that may be more deep-rooted than this, hence their presence in a wide range of early cultural belief.

To its critics, the structuralist's stress on how fundamental aspects of culture may be shaped around a relatively small number of basic rules is seen as locking early culture into a mesh of timeless structures. Yet for my argument, the way in which it highlights human culture's tendency to develop inertial forms, a point on which few would disagree, is precisely its significance. In particular, the extent to which anthropologists like Lévi-Strauss see the relevant principles of structure as hidden from the bearers of a culture, and only partially represented in any one cultural group, together with the assumed universality of the principles involved, all suggest that they sustain forms of cultural expression that are not susceptible to easy change. This is conveyed forcibly in Lévi-Strauss's general observation, one that embodies the very essence of his structuralism, that his approach enables 'the opposition between the collective nature of culture and its manifestations in the individual' to be overcome by treating the 'collective consciousness' as an 'expression, on the level of individual thought and behaviour, of certain time and space modalities of the universal laws which make up the unconscious activity of the mind' (Lévi-Strauss 1968: 65).

As this last quotation broadly hints, Lévi-Strauss's structuralism does not exclude time and change. Primitive society may seek to exclude change through ritual but the world is not always so obliging. In reality, he concedes that all societies 'are in history and change' (Lévi-Strauss 1966: 234). In fact, despite the charges against structuralism being timeless, his work has explored the relationship between structure and history, or synchrony and diachrony, at some length. It is a relationship which he sees as varying between different types of society, and especially between what he calls 'cold' and 'hot' societies. Whereas 'hot' societies are open to change, 'cold' or primitive societies, 'primitive for this reason', want to deny change and try 'to make the states of their development which they consider "prior" as permanent as possible' (ibid.: 234). In context, what this means is that the deep structures of 'cold' societies persist despite history. We can see this in his review of how classificatory schemes cope with change. Faced with fluxes in population, particular clans or moieties within a classificatory scheme might expand or die out. Groups work out solutions to these changes within the rules governing their system, so that change, the historical process, becomes 'a form without content' (ibid.: 235). When trying to theorize on the relation between structure and change, Lévi-Strauss drew on Saussure's distinction between languages that are more grammatical or motivated and those that are more lexical or arbitrary (Saussure 1983: esp. 65–98). When applied to other forms of cultural expression, he construes the shift away from the root order underlying culture as a shift from structures that have motivation to more arbitrary forms (ibid.: 156), though some might phrase it as a shift from structures that have a conceptual content to forms that have only an affective or sentimental meaning (Needham 1969: xxxvii).

Of course, primitive culture's overriding concern with persistence or inertia is a theme running through a wide variety of anthropological discourse. As a theme, it has been subjected to a powerful conceptual analysis in Eliade's *Myth of the Eternal Return* (1971). Primitive and archaic cultures constructed a celestial archetype of their world. This provided a model for how their actual world should be ordered. The two parts, their heavenly world and their real world, were linked via a point of centre, their *axis mundi*. In more advanced archaic societies and civilizations, like those of Mesopotamia and the Indus valley, the symbolism of this centre formed the basis for temples and even the first cities, a fact which has been developed by Wheatley in his work on the origins of urbanism (Wheatley 1969). Furthermore, a society's paradigmatic world, or celestial archetype, provided it with a model for how the geography of its real world should be organized. In effect, the world existed in theory or concept before it existed in fact, though we cannot really describe it as Platonic, as Eliade does (Eliade 1971: 34–5), given it was used as a blueprint for action. As a spiritual world that suffered no disasters or change, such a celestial archetype was bound to provide an enduring and unshakeable paradigm for earthly practice. In the process, the geography of the latter was arranged to fit the concept. For it not to square with its celestial archetype was to admit chaos or disorder. Their continuing complementarity was ensured by the continual recreation of their real world through ritual. In effect, the possibility of change, or 'the terror of history' (ibid.: chap. 4), was denied. Eliade's essay on aboriginal geography provides a well-attested illustration of how such forms of cosmography were as much process as form, but a process that was designed to repeat itself endlessly in an unchanging world. As with other cultures, the aboriginal world is seen as created according to a celestial archetype by culture heroes who awake from a sleep and move around their earthly world along dreamtracks creating all the main features and characteristics of their human geography as well as their physical geography. Periodically, the aborigines themselves follow the dreamtracks of their culture heroes, re-enacting the creation of their world, the cosmogonic act, through ritual in an effort to 'abolish profane time' and deny themselves the experience of history (Eliade 1971: 35; Head 1993: 481–99).

Eliade makes it abundantly clear that the ritualistic re-enactment of the cosmogonic act was a feature as much of archaic and traditional societies as of primitive ones. Each tries to deny change. Of course, in reality, the world is in a state of constant change. What particularly interested Eliade is how such societies coped with specific instances of change. In a sense, his observations here concern the fundamental question of how did human culture first come to terms with the reality of a changing world, or the problem of what he calls 'historical man' who 'consciously and voluntarily creates history' (ibid.: 141). His reflections on this history note two kinds of adjustment. The first invoked large cycles of creation/destruction/creation so that catastrophes and the like could be given a role within the cosmogonic system. The second incorporated critical events and personalities as metahistory, giving them mythological status. Clearly, both these types of adjustment coped by squeezing the insistent flow of history into the procrustean form of given concepts, reducing it to variations on a theme. For Eliade, and this was really the ultimate lesson of his study, the denial or suppression of time's flow continued down into relatively recent times. Only modern [wo]man is historical [wo]man, that is, only modern [wo]man is prepared to free [her]himself from the tyranny of *a priori* concepts of order and to allow the possibility of a history that brings real change because our concepts are opened up to genuine or critical reflection.

Eliade's conclusions can be compared with those developed more recently by Sahlins in his study of the impact which Captain Cook's arrival had on the South Sea islanders of the Pacific seen from the perspective of the latter. It is a study which addresses the same basic issue of how supposedly static cultural systems responded to the ongoing flow of history and events. Put simply, Sahlins emphasizes how the South Sea islanders absorbed their experience of Cook through their existing concepts and categories. In a telling phrase, he asserts that 'Captain Cook was a tradition for Hawaiians before he was a fact' (Sahlins 1987: 148). The point he is making is that reality is 'a dialogue between received categories and the perceived contexts, between cultural sense and practical reference' (ibid.: 144–5) so that the latter becomes a transform of the former. Hence, Cook was seen as a god. This is what Eliade was driving at when he tells us how archaic and traditional societies cope with figures who try to intrude historically by mythologizing them, making them a part of their cosmogony. However, Sahlins is taking us beyond this point. He wants to remove the opposition or distinction between synchrony and diachrony, stability and change (ibid.: 144). He does so by using a form of structuration theory. On the one hand, he argues that we 'know the world as logical instances of cultural classes' (ibid.: 146). But on the other, cultural concepts are always affected by the circumstances or situation in which they are used (ibid.: 147). In other words, the reproduction of culture through praxis invariably brings with it the possibility of its transformation. The precise relation between the two depends on the cultural order (ibid.: 152–3). Like Lévi-Strauss, Sahlins makes use of the idea that culture has an arbitrary side to its character, but he gives its meaning a slightly different inflection. As cultural systems move away from their received or core concepts through their experience of practical situations, so they become more arbitrary. For Sahlins, this means they become more historical, because they see the world in terms of their past, a past that cannot be read from the conditions of the present. Overall, what matters for Sahlins is his concern to soften the distinction between so-called 'cold' and 'hot' societies, by defining how the ongoing practical reference of all *a priori* schemes or received concepts ultimately makes for an arbitrary system.

Yet as Thomas (1991: 106–7) notes, in trying to elaborate on the conditions under which cultural order is open to arbitrariness or history, Sahlins shows how difficult it is to escape some form of oppositional distinction between societies that resist change and those that are responsive to it. Sahlins distinguishes between two broad types of society, one of which he labels as 'prescriptive' and the other as 'performative'. In the case of the former, societies seek to maintain their world in a steady state by means of the cosmogonic act, re-enacting its birth through ritual. In the case of the latter, they are open to change, responding positively to the demands of new circumstances. Yet his example of performative change hardly breaks free from the inertia which typifies any sort of pre-modern society. Describing the changes which affected Hawaiian society in the decades after Cook's visit, he talks about changes in customary concepts and relationships but 'always the functional revaluations appear as logical extensions of traditional conceptions' (Sahlins 1987: 140). Indeed, speaking generally, he cites Saussure's point that 'what predominates in all change is the persistence of the old substance' (ibid.: 153). Cook's voyages were of course sponsored by the Royal Society. As an instance of contact between a rapidly modernizing society and a traditional society, there are some doubts over how far we can take Sahlins's examination of Hawaiian reactions to it as a typical instance of a traditional society reacting to change. This quibble, though, does not alter the extent to which his argument casts pre-modern societies as inertial: they absorb change but without compromising prior concepts. In trying to bring such societies within a theory of change, rather than one which gives sole emphasis to their qualities of persistence, Sahlins is acknowledging that, overall, such societies do have a record of change and that what is at issue is how they negotiate the interaction between past and present.

Sahlins's study has similarities with Wagner's idea that culture passes through primary tropes or basic metaphors of meaning that each typify an epoch, a *longue durée*. Each primary trope constitutes a unit of self-reference, enabling each metaphor 'to expand the frame of its referentiality by a processual extension into a broader range of cultural relevance' (Wagner 1986: 9). In effect, the constitution of each trope or metaphor becomes a process, with primary or large-scale tropes being continually elaborated or re-expressed via a 'flow of analogy', each new analogy feeding the potential for still other forms. In the process, each new version serves as both 'proposition and resolution' in that 'it stands for itself' (ibid.: 15). However, the process of elaboration is not infinite. The potential of primary tropes can be exhausted.

In so far as Sahlins and Wagner see culture as an asymmetrical interaction between concepts and practical circumstances, their ideas have much in common with those developed by Bourdieu. As someone trained in the French anthropological tradition, Bourdieu's work displays themes and styles of interpretation that reflect the thought of writers like Durkheim and Mauss as well as Lévi-Strauss. Yet whilst there is a structuralist slant to what he has to say, his main position of argument, as in Sahlins's work, revolves around the assumption that culture is continually being opened up to change through practical circumstances and their history. In part, his differences may be a matter of context, for whilst he has worked with traditional societies, his philosophical statements are directed equally at modern societies. His work on the latter presents us with a structural configuration of society that is socially or culturally constituted. As with Giddens, this structure is not inaccessible, but one reproduced through 'strategies and practices via which agents temporalize themselves and make the time of the world' (Bourdieu 1992: 139). Again, like Giddens's rejection of 'mechanical dopes', Bourdieu's agents are 'not particles' but 'bearers of capital', or experience, who are able to negotiate their own position (ibid.: 108). In composition, his concept of society is fashioned around the twin concepts of habitus and field, the former being 'strategy-generating principles' that operate within people and provide them with the mental resources by which they can set about coping with the world (Bourdieu 1977: 72), whilst the latter represent the different sectors and institutional spaces across which they operate. Though habitus operates within human agents, it does not originate in the deep structures implied by the likes of Chomsky or Lévi-Strauss (Wacquant 1992: 18). Instead, habitus internalizes what Bourdieu calls the 'immanent law of structure' (Bourdieu 1992: 140), providing a mental template for the social structures or divisions that exist outside of it. Being the internalized mental template of a structure that finds expression through strategies and practices, it is open to being shaped by the latter. It is, in Bourdieu's words, an 'open system of dispositions that is constantly subjected to experiences' in ways that can reinforce or modify it (ibid.: 133 and 139). As Sahlins or Wagner might also argue, this openness is not at the expense of its capacity for inertia, being 'durable but not eternal' (ibid.: 133), hence his reference to generative principles being 'durably installed'. The problem for Bourdieu is that when one looks at work which explicates such ideas in context, notably his work on the Kabyle in Algeria (Bourdieu 1979: 133–53), his explication of habitus reveals generative principles that stand comparison with those found in wholly different cultures (e.g. Dodgshon 1985: 61–83; 1988: 69–82) and begs questions about whether some principles are 'durably installed' because they are more deeply structured within human agents than he might allow. This qualification apart though, what matters is that his work lends further support to the notion that all cultures are seen as establishing, or working through, large-scale structures of thought whose principles, if not their representations, respond only slowly to new or changed circumstances.

How writing and printing put geography in its place

The introduction of writing during the third millennium BC was an innovation so significant that its appearance is conventionally used as one of the key handful of traits by which we can typify the emergence of the early civilizations. It increased society's ability both to communicate and store information. As regards the former, it allowed for information flow between communicants who were separated both in time and space. Just as the development of language had allowed the symbolization of events and places other than those that were here and now, or immediate, so writing greatly extended the capacity for the flow of information to be displaced in time and space simply because of the way writing introduced an external form of data storage, a new form of memory (Lyle 1993: 63–4; Donald 1991: 333). It was now no longer necessary for the social interaction to be sustained solely by face-to-face contact, though primitive art had hitherto allowed some degree of temporal displacement because of its capacity for storing symbols. As regards storage, writing enabled more specific and structured information to be stored over time. Potentially, the past could now be visited on the present. A number of scholars have also noted how this increased ability to store and communicate information had implications for society's conceptualization of space. For Innis, it affected the geographical structuring of power, with durable forms of record (e.g. parchment, clay and stone) favouring centralization and hierarchical types of administrative order and those which are relatively impermanent (e.g. paper) favouring systems of spatial order that were more decentralized and less hierarchical (Innis 1972: 7).

Goody is equally sensitive to the effects which writing had on the integration of political and social processes over space, but he also draws out the extent to which writing added to the inertial properties of such systems and their associated beliefs. Two points lie at the heart of his case. He developed the first in an extended comparison of religion in pre-literate and literate societies. The former are conventionally seen as having myths, rituals and beliefs that are 'static, as persisting unchanged over the generations' (Goody 1986: 7). In contrast, his own reading of the problem in west Africa suggested that, as some forms of structuration theory would suppose, the oral transmission and re-enactment of rituals and beliefs involved so much shift in meaning through spontaneous invention or mis-statement that he sees them as flexible (ibid.: 7–9). The problem was that direct human recollection tended to be limited to about fifty years. Professional remembrancers could extend the record back further, but their styles of remembering had two consequences. First, they tended to alter information to suit the social context in which they were working, to validate it for a particular audience so that the 'remembered past was under constant pressure from the needs of the present' (Clanchy 1970: 166). The effect of this was not only to edit events and genealogies to serve the present but to compress time as it flowed towards the present and, in doing so, to suppress awareness of the passage of time (ibid.: 167). Second, the mnemonic devices used to store and recall data mentally were based on 'making language memorable' (ibid.: 169). To this end, extensive use was made of formulae and other mnemonic patterns such as proverbs, epithets, meta-phors and repetitions, or 'readily recallable thoughts' (Ong 1977: 47 and 285). Facilitating the latter was a tendency in oral cultures to structure thought around a limited number of themes that were absolutely central to the human lifeworld and its material survival (ibid.: 106). In effect, the system of informa-tion storage used affected the kind of information stored and even styles of thought.

For Goody, the introduction of writing, with its potential for casting ideas into a fixed and stable form of dogma, led to standardized forms of belief. Not only could ideas now be standardized across large areas of space but, with an original text always to refer back to, they could resist easy revision. He talks about the 'conserving power of written religions' (Goody 1986: 29), and goes on to suggest that societies with written religions were able to resist the colo-nial overtures of the West far more effectively than the more open nature of societies with religions or beliefs that depended on oral transmission (ibid.: 86). Writing also generated inertia through its capacity for storing and con-veying data about the basic facts of territorial order: administrative units, jurisdictions, tax burdens, landholding and so on, providing a benchmark of reference. For Ong, it also helped to standardize languages, particularly the mass languages, by giving them a fixed or defined form (Ong 1977: 40). Even more profoundly, it helped to objectify thought, by helping to distance the human mind from the word (ibid.: 47). Further, by codifying text in a geo-graphical way, positioning words on the page, it made possible the indexing of knowledge. In all sorts of ways, then, writing reduced arbitrariness and its capacity for dialect and the vernacular, with language use hardening into a received or standard form. In the process, it also standardized language in the landscape as well as on the page.

Writing has a peculiar double significance. It can be taken as a hallmark of early civilization and yet one whose eventual diffusion as a widely held skill was critical to the Modernity Project. The long interim phase was dominated by what are sometimes called customary societies, customary because their organization of relations rested on custom. In such societies, writing was an instrument of power, used to record obligations and rights (Larsen 1988: 178–88). In this respect, they were more insulated against change than the oral remembrance of obligations and rights. By the very nature of the written record, though, it admitted chronology. Time immemorial could now be reduced to the reign of particular rulers, and myth could start its transformation into history. From denying a world of change, however much they lived in one, societies could begin the process of temporizing their experience. From living in a world in which time was always ideologically if not actually reconstructed as a moment that continually repeated itself, they could begin the process of experiencing the flow of time as a Heraclitean experience, one in which nothing ever is and everything is becoming.

Inertia as a mirror to itself: the past in the present

For social theorists, the essence of the Modernity Project is the analysis of how the take-it-for-granted nature of the traditional lifeworld has been opened up in recent centuries to penetration by new stimuli. It is a shift that is best represented in a contrast drawn by Habermas. He uses work on the 'uncritical behaviour' of the Azande in west Africa in confining their thought 'more or less blindly to inherited interpretations' as his starting point, and contrasts their 'closed' worldview to the 'open' worldview adopted by recent cultures such as that of modern science (Habermas 1984: 61–2). The achievement of the Enlightenment in generating a more self-conscious and critical thought about both the natural world and the human condition can be seen as a primary mover of this more open worldview, one that has greatly enhanced the learning capacity of human culture. For Habermas, though, the emergence of more complex and specialized societies, with their hyper-differentiation of specialized roles, and their penetration or 'colonization' of the spaces of the lifeworld, has led to a legitimation crisis. The root of this crisis lies in the fact that the system needs of modern society have not been validated from within the lifeworld via a new normative consensus (Habermas 1987: 153–97).

Despite Habermas's talk of a legitimation crisis, though, the fact remains that when one excavates down into the lifeworld of modern culture, one is more surprised by what has survived than by what has been lost. Todd's work on the cultural basis of modern ideologies supports this (Todd 1985), as do Berry's thoughts on the cultural foundations of economic decision-making (Berry 1989: 1–18). It would seem that even after society has undergone the inversion from a condition in which economy is embedded in society to one in which society is embedded within economy, as Polanyi put it (Polanyi 1957: 57–75; Polanyi 1968: 158–73), there are still areas of the lifeworld that hold the market and its functional needs at bay. Indeed, much of the perceived tension that exists over the economic forces of global restructuring, between capital and community or market and place (Harvey 1989; Swyngedouw 1989: 32; Zukin 1991: 15), can be transposed into an attempt to distance system needs from those of the lifeworld and the struggles that surround the so-called cultural politics of place.

We also need to grasp how the enhanced learning capacity that has come with modernization has led to a more rationalized form of lifeworld, more rationalized not simply in regard to its part colonization by system needs and its heightened sense of the rationality by which society now functions, but also in regard to its normative consensus. Modernization may have produced a time–space compression in terms of the movement of people, goods and ideas, but it has also produced an equally significant time–space expansion in that, culturally, we are now more aware of time and space, as dimensions that have extension. Through the storage of records in written form, the development of new historical and archaeological methods by which the meaning of time can be recovered and, above all, through a more reflexive approach to time, there has emerged a new sense of history as the chronology of human experience and change. Likewise, through exploration and a new ease of mobility, and through a vastly increased level of spatial and global interaction, there has emerged a new sense of geography as a map of global differences. We can draw a revealing contrast between this new awareness of time and space and that experienced by early cultures. Modern culture has greatly expanded the idea of time, drawing out the flat stationary concepts of early culture into a deep chronologically based sense of time as history. Local territories that were totalized as a worldview by early cultures have now been revealed by travel and interaction as but a small part of an infinitely larger world. In effect, what was thought to be an experience of the geographical world in its totality has been re-proportioned into what is simply local. Likewise, the equation of the present, or the cycle of the year, to time in its entirety has been exploded, with time being greatly enlarged so that it can now be seen to stretch backwards and forwards to an infinite degree. Further, its flow has become an increasingly measured and costed experience (Thrift 1981: 56–84). These are profoundly meaningful shifts in the experiential world of human agents, changing the world as they perceive and experience it from within.

I want to argue that such changes have affected how we relate to the world around us. Whilst the rise of the global economy and the relentless spread of free markets has threatened to overwhelm space everywhere with the singular needs of capital, to configure it around the forms of a single world system, shifts in the temporal and spatial references of the lifeworld have spawned alternative evaluations of space and time. Indeed, one could argue that whilst there has patently been a shift in the way space is evaluated and ordinated, and in the scale at which the different systems operated, pre-capitalist systems were no less capable of applying singular principles of order. I say singular rather than uniform because there were processes within pre-capitalist no less than capitalist societies that were capable of differentiating space into core and periphery, both in an absolute and a relative sense. Seen in this way, the shift from standardized or indoctrinated forms of intersubjectivity towards a more self-rationalized lifeworld, one informed not just by time and space but by history and geography, becomes significant. The shift can be measured in the way that differences in time and space, the textures of history and geography, are now valued. Everything from the past and everything that gives identity to place now has a value of sorts to somebody, somewhere. What was once seen as the 'deadly disease of nostalgia' (Lowenthal 1975: 1) is now seen as therapeutic rather than threatening. It is the differences between particular times, or particular spaces, not their equivalences, that now matter. Their impact on landscape inertia is far-reaching. The post-industrial landscape has become overlain by restrictions and management regulations laid down by a complex web of landscape managers, from environmental protection agencies to civic conservation societies, from local authority planning groups to heritage landscape trusts. The old jibe about Italy being a nation of museum keepers is now far more widely applicable as developed societies everywhere try to freeze the landscape, to make a virtue out of inertia. As Lowenthal once said, 'to appreciate the past is to transform it' (Lowenthal 1979: 125).

These shifts in the way human culture values differences in time and space can be brought together in the way society now symbolizes place, attributing a precise geographical and historical value to it, one bound up with its identity. From the time when societies saw their *axis mundae* as the symbolic centre of their world, they have exploited the symbolic capital that can be invested in place. As Duncan put it in his study of the Kandyan kingdom, the human landscape becomes 'an ordered assemblage of objects, a text' which 'acts as a signifying system through which a social system is communicated, reproduced, experienced, and explored' (Duncan 1990: 17). In its role as text, as a source of what Duncan refers to as 'rhetoric', landscape can be used to signify many fundamental attributes of a culture. Duncan's study was concerned with reading the relations of power that existed within Kandyan society. Other studies are less explicit about the symbolic text written into landscape but are no less supportive. In Europe, we see it during the medieval period in the way the Florentines 'were learning how to look: and as the built environment came within their purview they enthusiastically set about to enhance its beauty in accord with their pride in the state' (Goldthwaite 1980: 76). Likewise, when the area surrounding the centre of Vienna was replanned in the mid-nineteenth century, the opportunity was taken to signify the new political order that had just come into being. The walls partly encircling its old inner city had, for strategic reasons, been surrounded by dead space. Following the Revolution of 1848, plans were commissioned to fill this semi-circle of undeveloped land as part of a city expansion scheme. The plans were published in 1859. Instead of using the area to link the old city – with its symbols of pre-1848 Austria (e.g. the Hofburg or Emperor's palace) – to the outer suburbs, the plan surrounded the inner city with a self-contained ring of boulevards and gardens, the Ringstrasse. The effect was to separate the old from the new (Schorske 1981: 32). With equal meaning, the Ringstrasse acquired the token symbols of the new order: the Parliament, Rathaus, the University and the Burgtheater. Though the classicism of the Ringstrasse was being heavily criticized before it was even complete, its association with a new order of things imbued it with more than mere design or bricks and mortar. It signalled a new direction in Austrian political thought and, for that very reason, had a potential for inertia built into it. In short, it provided, to use Meinig's phrase, an 'iconography of nationhood' for a new Austria (Meinig 1979: 164).

Society's capacity to symbolize its cultural landscape in ways that inhibit easy change has not diminished over this century. However, the increasing shift towards a more rationalized lifeworld has brought about a shift in how such landscapes are construed. Put simply, the symbolic landscapes created by early culture were inertial in their effect because they were brought within stable, enduring belief-systems that were shared by all and treated not as something past or in change, but as stable or constant expressions of the present, the moment in being. With modern societies, landscapes are increasingly being symbolized precisely because they are deemed to be inherited from the past or traditional, that is, as having inertial qualities. We are reminded of Bourdieu's point that 'symbolic power is the power of consecration and revelation, the power to conceal or reveal, things that are already there' (Bourdieu 1990: 38). Modern societies choose more and more to consecrate and reveal what is past, what is in time. However, there is nothing uniform about their response. To return to the point made earlier, what is increasingly significant is the way in which time and space have acquired a heightened significance through awareness and appreciation of difference, that is, through the history and geography written into them. This decentralization of normative values within the lifeworld means that what typifies modern society is the way everything has a value for someone and the way, as a consequence, there is a latent pressure to preserve or conserve everything. Looked at from within, it is a selective but coherent world. Looked at from without, or at a societal level, it is a fragmented world whose parts seem divided by historical fault-lines and uncomformities. Indeed, misleadingly in my view, some would extend this sort of point. They have argued that the present is a disjunctural moment and, because of the way the message of the past is now so fragmented, modern society has become cut off from its real past. The 'past as referent finds itself gradually bracketed, and then effaced altogether' proclaims Jameson (1991: 18). In fact, the opposite is true.

Arguably, the idea of the past as a mosaic of fragmented images has come about through the scale of recent development in modern societies and through the way in which such societies have increasingly objectified their historical sense of place, separating out relict landscapes and their features as exemplars or signifiers of the different phases that make up what is an increasingly historized past of each place. Concerned at the modern tendency to symbolize the past by preserving it, Lowenthal suggested that to objectify the past like this is to separate it out from the present as something different, to separate past things 'from continuity with the world around them' (Lowenthal 1979: 112). This treatment of the past for its own sake, that is, as a source of history, has the effect, he argues, of altering it, of making it different from what we think it is (ibid.: 124; see also Lowenthal 1975: 27). In fact, he himself calls for a treatment of the past more in keeping with that held by traditional societies, one which treats what has survived from the past as part of the present. 'We require a heritage', he argued, 'with which we continually interact, one which fuses past with present' (Lowenthal 1985: 410). The distinction is between a past which is consciously preserved as a field-record of defined historical forms and a 'fixed past' which is unconsciously preserved by being 'assimilated in ourselves, and resurrected into an ever-changing present' (ibid.: 412). In actuality, as Lowenthal would be the first to acknowledge, both forms of past, and both forms of inertia, exist side by side. Furthermore, however much we argue their differences as ways of relating to the past, there is a sense in which the growing commodification of such landscapes by the heritage industry threatens to reduce their 'historical differences to the realm of the same' (Crang 1994: 344), but this does not alter my basic point.

The emergence of literate societies coupled with the spread of a more objectified and critical understanding of the past are not the only factors at work. Some would also see the growing interest in preserving the past by preserving place as a response to the homogenizing pressures of globalization and the blanket of anonymity which has been cast over urban centres in widely differing cultural contexts. Traditional or indigenous landscapes have become a focus for preservation movements simply because they are bound up with notions of self-identity in a world in which difference is being rapidly eroded. Where prominent symbols of the local and global, the indigenous and the extra-local, are juxtaposed in close proximity, some have argued for seeing such landscapes as contested landscapes (e.g. Ley 1987: 40–56). In essence, it is also a contestation between the 'past in the present' and the needs of change. Yet some might argue and rightly so that there is a great deal of cross-dressing in the making of such landscape differences, with many new centres of investment providing an arena for post-modernist expression and the extensive but eclectic borrowing of symbols and design features from the past. Images of past architectural styles, as well as speculations about future styles, are quoted in a fragmentary form or used in a wholesale fashion to create a false sense of a different time and place. Such illusions of form are seen as contributing to the superficiality with which the forces of globalization intrude and transform local space, crudely displacing the locally felt experience of time and space. Yet the fragmentary nature of post-modernist architecture and the anachronistic images and landscapes which result from it are really no different in their effect from the mosaic of preserved spaces and meanings that have resulted from local preservation movements. Both need to be seen as part of what we can call the indexing of knowledge. As Ong has said of the first printed book, once the written word was standardized on a page by page basis, then data could be recalled via an index. Increasingly the indexing of knowledge, with data retrieved systematically rather than in its original composite or contextual form, has become a feature of data storage and retrieval. Similarly, selectively preserved or revalued, the landscape of the real world now becomes part of the jumbled way in which information or experience is increasingly stored, with order or meaning being restored through the way it is recalled and symbolized. The latter is an act of mind, as it was for early culture.

Admittedly, significant differences still exist between societies in this respect. Lowenthal, for instance, talked about how the stock images of American society depict it as 'neglectful of the past, impatient of the present, engrossed only in the future' (Lowenthal 1976: 89). By comparison, 'delight in continuity and cumulation is integral to English appreciation of *genius loci*, the enduring idiosyncracies that lend places their precious identity' (Lowenthal 1985: xviii). Berry's comparison between urban planning in America and Canada confirms that cultural differences in approach do exist, and can affect the degree of inertia that we can expect to find in landscape (Berry 1989: 5–6). Yet others would now read recent American society differently. Hayden's call for the conscious use of 'place memory' within American cities as a way of connecting the experience of the past to the present demonstrates this (Hayden 1995: 11 and 78).

Contrary to what might be assumed, early cultures did not necessarily establish conditions for inertia through fixed or stable adaptations with their environment. Unless they were locked into specific strategies or solutions by heavy investments of energy that accumulated over time, early cultures had too many resources of flexibility to remain static despite environmental change or population growth. A more likely source of inertia for early cultures was their ideology and the categorical order provided by their concept of cosmography. These ideological constructions of the universe provide a paradigmatic model for the ordering of their real world. The concept runs ahead of the practice, with their relationship being continually re-established through ritual and myth. In such a world, the idea of change is denied, even if reality has to be more pragmatic in coping with novelty and change. If early culture fostered inertia by not thinking about change, modern societies have fostered symbolic or cultural sources of inertia by doing just that. Through the opening up of the lifeworld to reflection, along with the enhanced learning capacities of modern societies, there has been a dramatic time–space expansion. The culture of modern society is more aware of time and space, more aware of its different moments and spaces, appreciating how the here and now is only one moment in a succession of moments that are different or have history, and appreciating how their local worldview is only part of a much larger universe of spaces which, like moments in time, are distinguishable by their differences. This more objective approach to time and space has unfolded during the very period in history when the world has been progressively integrated around a single world system of trade and economy, a world system that has pushed the world towards an opposite trend of time–space compression as regards movement, interaction and difference. In such circumstances, it was inevitable that we should be witnessing a cultural reaction in which perceived differences of time and space, or the history and geography inscribed in particular places, should be preserved against the levelling effects of globalization and against the widespread and place-insensitive changes wrought by the global economy's need for regular restructuring. As with early culture, though, the inertial qualities of this reaction have taken on a moral force in the obligation or duty to preserve, making it a matter of cultural politics.

6

Organizational sources of inertia

Institutions are like fortresses. They must be well designed and properly manned.

(Karl Popper, cited in Johnson 1966: 88)

We can define the institutional or organizational forms of a society as being made up of either integrated, unified systems of roles, rules and relationships or as focused networks of information gathering, processing and decision-making. These different dimensions are relevant to the definition of societal organizations throughout history, but over this century, the memory-management aspects of their character, their ability to store, handle and communicate information, have come more and more to the fore (Lash and Urry 1994: 320). Yet no matter how we strike the balance in the way we define them, the fact remains that they form the constitutive core of society. To suggest that these organizational forms are a powerful source of inertia is hardly novel. Almost by definition, they are an aspect of society's character which is least responsive to change and which has a delayed relaxation time when faced with pressures for change. Almost as soon as they come into being, they have a tendency to become inertial and to act as a constraint on any pressures for change or for new forms. Indeed, academic studies of their basic character invariably stress their resilience to change.

In part, such a view owes much to the fact that the organizational forms of a society are a prime means through which it acquires and sustains structure. In reply, some would argue that since such structures are subsumed within process, continually reconstituted through everyday social practice, the latent inertia embodied in structure is easily overstated. This sort of counter-view, though, tends to ignore the asymmetrical relationship between everyday practice and structure, with the former made up of particular instances, of individual agents and their roles, and the latter of aggregates. Arguably, the inertia attached to organizational forms resides in this asymmetry and the fact that one articulates only part of the other. It does not encompass it. Part of their inertia, though, can also be attributed to the assumption that the viability of a society, its ability to endure in a stable form, depends on the possession of order which, in turn, depends on the effectiveness and coherence of its organizational forms and how well these forms map the normative goals of a society. The one engenders the other. As the quotation by Popper used at the head of this chapter makes clear, once the organizational and institutional forms of a society are seen not just as embodying its normative values, but as protecting them, then we can expect such forms to be a powerful source of inertia.

This last point of argument is a recurrent theme in the debate over functionalism, especially structural functionalism. Put crudely, structural functionalism defines the essence of societal systems in organizational terms. It sees them as structured around sub-systems or institutions. Each performs a specialized role or function, usually within a fixed and defined spatial context, so that the different parts only have meaning and viability when seen in the context of the system as a whole. Collectively the different sub-systems and institutions are seen as co-ordinated through their goal behaviour, each contributing to a common set of system goals. This is especially true of what some have called normative functionalism, with its stress on value co-ordination as the tie-string of the system and the prime source of stability. Not surprisingly, criticism of structural functionalism has focused on the fact that it offers a theory of society through a theory of persistence, one that concentrates 'attention on the stabilizing, the pattern-maintaining or the boundary-defining processes in social life' (Bock 1963: 232). In effect, it painted a picture of society at rest. By doing so, it created a dilemma for its proponents. If societies are organized for stability, how does change come about?

Change and the flexible organization

The challenge of bringing change into an organizational and institutional perspective has been addressed in a variety of ways. For those who favour a standard organizational approach, the solution was either to introduce external or environmental factors to disrupt society's steady state or to modify the concept of functionalism in a way that allowed for organizational change. In the case of the former, otherwise stable social systems are seen as falling foul of unfavourable conjunctions, colliding with each other during phases of rapid growth, or being undermined by changes in their environment (e.g. climatic change, volcanic eruptions). In the case of the latter, change is seen as unfolding smoothly from within through differentiation. Talcot Parsons's work can be used to illustrate this change-through-elaboration thesis. In its developed form, Parsons's model organized society around a hierarchy of specialized sub-systems based on the broad functions of pattern maintenance, integration, polity and economy (Parsons 1966: 25). Over time, each of these sub-systems copes with growth by becoming more differentiated, splitting into more and more specialized roles. As they do so, the system's value pattern has to be framed in more generalized terms so as to legitimate their more diverse goals and functions (ibid.: 23). Though he refers to the 'wider variety of goals' here, he really means the more detailed specification of goals, one that mirrors the more detailed differentiation of functions. In essence, differentiation is a process of change from within, one that elaborates the structure of society around a given theme or goal. Though some have tried to portray Parsons's modifications to functionalism as converting it into a neo-evolutionary model (e.g. Smith 1973), one could argue that he has really converted it into a developmental model, one that plays out the finite potential locked within a particular formative framework.

An alternative way of explaining how the organizational forms of society are more amenable to change is through some form of structuration. In essence, structuration removes the structural independence of social organizations and institutions, their supposed super-organic qualities, by seeing them as reproduced through social practice. Carried forward only through their embodiment in social practice, their character is thereby made accessible to human agency. The two are seen as interacting recursively, each capable in the right circumstances of determining or influencing the other. I have no quibble with this as a general statement of their interaction and as a statement of how the individual and the collective are linked. It acknowledges that there are circumstances in which people are empowered to determine organizational change and other circumstances in which they may be constrained by the structures which they have created. It sees these different circumstances, though, as linked solely in a successive, recursive way.

I would distance my own position from this in three ways. First, for reasons that I have already touched upon in my opening paragraph to this chapter, I would argue that whilst this allows constraint, nevertheless, it still understates the part which the organizational forms of society can play in constraining social practice, in being the dominant partner, simply because it does not allow for their basic asymmetry. Many of the organizational forms that make up mature societies comprise rich archaeologies of past experience, accreted layer by layer over time. They exist in the present but are hardly contemporary with it. The current debate over legacy programs is part illustration of this, with the routines and practices of many large organizations seemingly organized around what in many cases are literally thousands of computer programs that have accumulated in a piecemeal and accretionary way over the past three or four decades and which now pose huge problems and costs in negotiating the brief step from 1999 to 2000. Second, I think we need to be much clearer about how social practice reproduces organizational structures. It has become a commonplace to use the biological analogue of reproduction, yet whilst the comparison has some validity, it carries meanings which have not been properly explored. Biological reproduction is a process determined genetically. By contrast, social 'reproduction' is a more disparate process of acculturation, conditioning and learning. Much of what is acquired through acculturation and conditioning is part of an inherited take-it-for-granted world. For Whitehead, 'the number of important operations which we can perform without thinking about them' underpins the very advance of civilization itself (Whitehead 1939: 61). By comparison, those concepts and information that we deploy more thoughtfully via day-to-day practice are, by their very nature, more open to easy revision. Indeed, some would see such concepts and information as in a different mental category to those used repeatedly 'without thoughtful inspection' (Bateson 1972: 509; see also Schrecker 1948: 181–203). In effect, therefore, we need to distinguish between different dimensions of social reproduction. Different scales are involved, rather as one might distinguish between changes in the information that is stored in language as a system of grammar, as a system of codes and meanings, and that which language is actually used to communicate. Such changes patently occur at different scales. We cannot meaningfully say that they are all locked recursively into the same relationship between everyday social practice and structure. Thirdly, we need to consider whether we can more closely specify the practical circumstances, the different geographies, under which social practice can affect the structural organization of a society and vice versa.

This last point is a question which Unger has posed. Unger's thesis is based on an analysis of institutional forms which contrasts 'the pull of plasticity' with 'the push of sequence' (Unger 1987c: 212). It seeks a general understanding of change by rejecting the idea of integrated holistic systems of society that are replaced *in toto* by other ready-fashioned total systems. Instead, it envisages a pattern of piecemeal replacement leading ultimately to cumulative change. A key concept of his argument is the notion of formative context, a context that has both institutional and imaginative dimensions (Unger 1987b: 58). It is formative because it shapes social routines, roles and hierarchies in the spheres of politics, exchange and warfare (Unger 1987a: 155; 1987b: 58). In composition, a formative context is made up of a set or combination of institutional arrangements. These sets or combinations are not given or indivisible. Indeed, it is of the essence to his case that in theory any number of combinations is possible, though in actuality constraints operate to restrict the actual number (Unger 1987a: 91; 1987b: 170). The problem for society is that, once developed, formative contexts tend towards entrenchment and rigidity. They generate and sustain 'a richly developed set of practical and imaginative routines' but, in doing so, they have 'a corresponding tendency to resist disturbance' (Unger 1987a: 88–9). In part, this tendency develops from the 'push of sequence' or the past. 'Each institutional and imaginative order', Unger maintains, 'influences its sequel by giving a bias to the outcome of order-transforming conflicts' (Unger 1987b: 250).

Defined as such, the notion of a formative context provides Unger with a context not just for stability but also for change. It requires us to understand how context-preserving routines become context transforming. Central to his case here is the notion of what he variously calls disentrenchment, denaturalization or false necessity (ibid.: 278). Disentrenchment is straightforward and simply means the extent to which institutions and their values can be disembedded from their prior context and recombined in new ways. Denaturalization is the extent to which such institutions 'can be challenged in the midst of ordinary social activity' (ibid.: 167), so that the gap between what is seen as preserving such roles and their hierarchies and what is seen as challenging them narrows. Less entrenched and denaturalized contexts lessen the grip of rigid roles and hierarchies upon new forms of collaboration, 'bringing framework transforming conflict and framework preserving routine closer together' (ibid.: 249). Circumstances of plasticity are thereby engendered whereby new institutional combinations can be tried out and new formative contexts evolved. The extent to which institutions can be disentrenched or denaturalized so as to create these conditions depends on 'the pull of negative capability'. This is an idea rooted in concepts of ecological adaptation and the notion that well-adapted systems, with a diversity of specialized sub-systems, have greater difficulties changing than more generalized systems. The redundancy displayed by more generalized systems provides them with what Bateson called 'unused freedom' (Bateson 1972: 285), that is, greater amounts of disentrenched or denaturalized social space and resources. Put in more direct terms, what Unger is saying is that disentrenched or denaturalized systems are less constrained by the inertia of existing structures.

There is much in Unger's discussion that has relevance to my own case. In particular, I want to underline two points of argument. First, there is his stress on how the structure or organization of societal systems is such as to leave it open to change not as an integral or indivisible system but in a differential way, with institutions being more capable or susceptible of disentrenchment and denaturalization in some contexts than others. Second, there is the way in which he tries to explain this differential openness to change, with its stress on the relative entrenchment or plasticity of institutional combinations. Though he discusses geographical instances of change in these terms, such as those of Germany and post-revolutionary Japan, there is no sense in which he offers his argument as a geographical analysis of change. For Unger, such examples illustrate a potential for non-revolutionary change that resides in all societies. In other words, his argument differs only in means but not effects from that offered by those like Buckley who likewise talk in terms of societal systems or organizational forms as revisable. Yet as I will argue at greater length later, Unger's ideas on relative entrenchment, denaturalization and plasticity can be given far more explanatory power if seen as entrenched or disentrenched in a geographical sense. If this is to be done, we need a sharper recognition of the geographically based distinction between the conditions under which institutional forms are entrenched as opposed to those under which they are open to revision.

Change and the inflexible organization

Unger offers a bridgehead to such ideas through his stress on the tendency of formative contexts to incline towards rigidity. For some, this cul-de-sac of development is less easily avoided than Unger would have us believe. In some formulations, there is no case to answer. The social philosopher Nisbet, for instance, has argued that far from being a weakness of functionalism, stressing society's avoidance of change, its capacity for equilibrium and stability is actually a source of strength for it as an interpretation (Nisbet 1969: 283). For society, change is 'not natural, not normal, much less ubiquitous and constant. Fixity is . . . in place and in time, we find over and over that persistence in time is the far more common condition of things' (ibid.: 270). Change for Nisbet arises externally, through the conjunction or collision of systems (ibid.: 281).

Nisbet's review of the problem amounts to a general survey of social norms and values in so far as they service the basic inertia of societal organization. Others have addressed the problem through more specific analyses. Of particular interest are those who have couched the problem in evolutionary or developmental terms. As a group, these are largely based on prehistoric or historic case studies. Collectively, they introduce a variety of ideas into the debate. Johnson's work on early civilizations of the Near East during the third and second millennia BC considered the problem of how the organizational structure of society was affected by the fact that particular channels of communication and information exchange had finite capacities which, if reached, produced overload and reduced efficiency. Any sustained scalar expansion needed to introduce successively higher levels of control, selectively filtering information and decision-making, in order to cope with the increased volume of information that had to be processed. What Johnson proposes is that as societies approached this threshold, their existing structures experienced a downturn in efficiency, greater costs and growing dysfunctions. Only those which innovated by adopting better methods of information storage and transmission (i.e. writing), and higher levels of control, selectively filtering the upward flow of information and creating a still more hierarchical system or centralized system of decision-making, survived (Johnson 1982: 389–421). This same problem of how early civilizations coped with the emergence of complexity was also considered by Flannery. He centred his analysis around the twin processes of what he calls linearization and specialization, with linkages between the main components of the system being organized so as to emphasize the centralized flow of data handling, both vertically and horizontally, through the selection and promotion of key centres and nodes. Having defined how change leads to more complex forms of organization though, he then introduces constraints and limitations, with sub-systems being seen as ultimately moving from a state of system-serving to self-serving (Flannery 1976: 117).

Reviewing organizational changes in the same sort of context, or early civilizations, what impressed Tainter was the extent to which many ultimately either collapsed or stagnated. He explained this lack of sustained growth and change in terms of the costs attached to particular forms of organization. An 'immutable fact of social evolution' is that as society evolves towards more complex forms, 'the support costs levied on each individual will also rise, so that the population as a whole must allocate increasing portions of its energy budget to maintaining organizational institutions' (Tainter 1988: 92). In effect, social organizations are subject to the law of diminishing returns (cf. Boulding 1971: 29–30). Up to a point, they successfully convert energy – as measured by investments in agriculture, crafts, education and so on – into socio-political organization. Initially, the cost-benefits of such investments are favourable. However, sooner or later, a point or threshold is reached at which further units of investment do not yield a proportional return, but instead, yield smaller and smaller increments of return on investment. For Tainter, such increases in socio-political organization have an exponential effect on costs so that the return on further energy inputs into organization are not commensurate. When societies reach this point, they are faced with stagnation or collapse. His own conclusion is that where a collapse back to simpler forms took place, it represented a rational solution to the problem of rising costs, enabling societies to return to an efficient level of energy use in relation to their socio-political organization (Tainter 1988: 121).

The same problem of why organizational change is not sustained has been explored in a historical context by the economist Olson. His argument is straightforward. Organizations and collusions for collective action, such as medieval craft guilds, 'usually emerge only in favourable circumstances' (Olson 1982: 40). Once present, though, and given a stable society, they become numerous. In addition, they became more complex in their behaviour, with more bargaining, more elaborate arrangements over organizational activity and more work rules (ibid.: 71–2). At the same time, 'custom and habit play a larger role' (ibid.: 86). Their shift towards more complex forms is accompanied by a change in the goal behaviour of groups. Instead of seeking higher levels of output, they increasingly organize themselves as distributional coalitions working to secure for themselves a larger share of national output, inhibiting society's willingness to adopt new technologies and to change how resources are allocated in response to changing conditions. The sum effect is a reduced rate of economic growth (ibid.: 65). What Olson is offering us is a theory of inertia founded on the character of socio-economic organizations and their collusive nature: the longer a society is stable, the longer it has stable boundaries and the longer it enjoys freedom of organization, 'the more growth-retarding organizations it will accumulate' and 'the more it will positively resist change' (ibid.: 125). Though directed specifically at the way organizations bargain over their economic position in the context of the marketplace, Olson's argument ranges widely, drawing heavily on their use of socio-political power in securing their position and in defending it against change in market conditions.

Olson's stress on the potential inertia surrounding the behaviour of socio-economic organizations was developed through historical examples but can be matched by a wide-ranging debate that uses present-day examples to make what is essentially the same point. A notable contribution has been made by Arrow. His work has concentrated on modern economic organizations as systems designed for the handling of information. In relation to the argument that I am trying to construct here, two of his propositions are particularly relevant. The first concerns the economics of organizations. In essence, organizations are designed to 'take advantage of the productivity of joint actions' (Arrow 1974: 53). However, there can be no gains in the productivity of information handling if all information is relayed to everyone else in the organization. In consequence, and at the very root of their growth, there is a tendency in all organizations to increase productivity through the selective or specialized handling of information, through the greater co-ordination of the sub-units handling these different types of information and through the introduction of specialized codes to facilitate information storage and communication (ibid.: 37 and 53–5). This inevitably leads, as Johnson and Flannery reasoned in a quite different context, to organizations that are both more unitary and structured in design. Conceptually, we can see this shift towards more hierarchical forms in terms of the AR:PR, or the ratio of actual relations to the potential number of relations. Higher order structures gain efficiency and effectiveness by increasing the AR:PR ratio (Hallpike 1986: 245–52). In the process, organizations become more adapted to a given set of needs and opportunities. Being more adapted to a given set of needs and opportunities, mapping them into their structure, is a feature of all organizations, not just those which are economic. However, its cost is the loss of flexibility.

The second feature of Arrow's argument concerns the learning cost of organizations. Complex organizations involve specialized roles and codes which, 'being locked up in an individual's mind, [are] necessarily irreversible' (ibid.: 40). What he means is not that the individual cannot learn new codes but that the energy invested in learning one code cannot be converted via some energy translator into another. What has been invested in a code, and in all such knowledge, is largely written off. As well as being an irreversible commitment for the individual, it is also a heavy and 'irreversible capital accumulation for the organization' (ibid.: 55). The roles and codes generated by organization, though, have one further implication. Their 'history matters' (ibid.: 56). Their roles and codes are fixed in accordance with the 'best expectations' at the time of the organization's creation. More to the point, once an investment has been made in particular information channels, it will be cheaper to go on using them than to replace them with new ones simply because of their heavy cost (ibid.: 41 and 49). The sum effect is one of inertia, with organizations locked into particular forms and chronologies of irreversible investment unless they are prepared and able to write off such capital forms and start again.

A comparable study of the economics of organization that has lessons for us has been presented by Nelson and Winter in their evolutionary perspective on economics. Like Arrow, they see organizations as ultimately 'reducible to the knowledge of [their] individual members' (Nelson and Winter 1982: 104). However, they elaborate further on how this knowledge shapes organizations in a distinct and revealing way. Organizations, they argue, have a memory. This operational memory consists of their routinized activity, a routinized activity that is developed around specialized roles, skills and codes, the latter including 'a generalized language competence' and knowledge of the 'organizational dialect' (ibid.: 104). But whilst part of this operational memory consists of knowledge that is explicitly understood by its members, a significant part 'is remembered by doing' (ibid.: 76 and 134). This distinction between the operating routines of an organization that are explicitly understood and those that are 'remembered by doing', between the knowledge which its members can articulate and that which is tacit, echoes my earlier point about differentiating between the various ways in which social knowledge is reproduced. It has a bearing on how Nelson and Winter interpret the responsiveness of organizations. Seeing the problem in terms of firms, they go on to argue that so much is tacit, so much resides in the organization's memory, that we cannot expect flexibility to characterize their behaviour (ibid.: 134). As a working rule, we can expect firms 'to behave in the future according to the routines they have employed in the past' (ibid.: 134).

Nelson and Winter were primarily concerned with the organization of the firm. Given the standard definition of the firm as an attempt to internalize sections of the market in which it operates, their ideas can be profitably applied to the organizational forms embodied in markets generally and to regional economies. Over recent years, there has been a renewed debate amongst economic geographers on the nature of industrial complexes and regions, and the organizational linkages that exist beyond the individual firm. Initial attempts to see industry in these wider terms concentrated on some form of agglomeration theory, with its stress on the external economies that competing firms can derive from clustering in the same locality. As Scott put it in a review of thinking on industrial complexes, just 'as single firms expand or contract both along the dimensions of scale or scope, so complexes of firms expand or contract both along the vertical and along the horizontal axes of production organization', thereby bringing 'into play a series of important system effects' (Scott 1988: 28; see also, Walker 1985: 248). These system effects range over such gains as the cost reductions to be drawn from having a local concentration of information about competing suppliers and services, and the benefits to be derived from sharing basic infrastructural facilities. As Scott himself makes clear, more recent analyses of industrial complexes have widened the debate in a significant way. In addition to seeing regional complexes in terms of production linkages or in terms of the services and infrastructure which they share and draw cost-benefits from doing so, some now stress the extent to which an industrial complex has to be seen in terms of its wider socio-political organization. Particular concentrations of industry become associated with particular skill pools, labour routines, levels of living, union and political activities, place-specific cultural practices and institutions and so on (Storper 1985: 274 and 278; Scott 1988: 12). Construed in these wider terms, the organizational complexity of industrial regions takes on a much richer meaning.

We can add to this enlarged meaning, though, a further point of argument. Not only is there more to the organizational basis of industrial complexes than production linkages, but all the different dimensions – production systems, skill pools, labour routines, information channels, codes, etc. – tend to be brought together and co-ordinated through particular phases of investment. Collectively, they represent a unique mesh of linkages and orientations forged during particular phases of change (Massey 1984; Gertler 1984: 172). Some have tried to capture this time-bound and organizationally constrained nature of local economies and their societies by using terms like locality. Clearly, it is a boundedness that has a bearing on how they cope with subsequent change. Once localities are underpinned by an economic and cultural stratigraphy of successive forms, change itself becomes conditional. Firms and individuals alike are presented with the challenge and risks of disengaging from an established order of things whose organizational linkages and ramifications extend far beyond the bounds of the firm. Given the fixed orientation of so much of the organizational behaviour and investment involved, we can hardly be surprised if mature industrial regions appear burdened with lags and inertia, with their 'territorial organizational framework' and 'weight of tradition' sustaining them in a 'self reinforcing way' (Storper and Scott 1989a: 145–6). Irrespective of the pressures generated by declining rates of interest and capital migration, a powerful reason why industrial complexes have a tendency to persist with old forms is because they keep on working organizationally, the orientation of each part or link being sustained by the very selective external expectations placed on it by adjacent or linked parts of the organization. Seen in this context, Scott's observation that an individual firm cannot 'unilaterally control the social conditions of its own existence' (Scott 1988: 29) clearly has considerable implications. It helps us understand how most forms of organizational inertia are geographically emplaced.

So far, I have dealt with the organizational inertia that has been read from socio-economic organizations by economists and economic geographers. However, the basic principles of what they have to say can be extended to other types of organization and their susceptibility to inertia. For example, despite their symbolic role at the heart of revolutionary theses of change, political organizations are especially prone to inertia and the constraining hand of prior structures and routines. It is a theme developed in Hocart's analysis of how the organizational forms surounding early tribal chiefdoms and kingships persisted in the structuring of power and roles around early state rulers (Hocart 1970). Likewise, as I have tried to show in chapter 4, the persistence of political forms is a prominent theme in the subsequent history of state systems. Nor are modern political or administrative organizations exempt from the problem of persistence. This is well shown in Kaufman's analysis of political organizations. Speaking specifically about political organizations, he argues that they 'are born in large numbers and die in large numbers' (Kaufman 1975: 133). Yet whether they last a few days or for centuries, he posits 'they are incapable of changing their structure or behaviour in any but the smallest degree' and that their survival rate and success depends on 'chance variations in their relatively inflexible behaviour and chance characteristics of the local environment in which they happen to occur' (ibid.: 133). Later in the same discussion, he again stresses the importance of environmental flux to their survival. Whether their 'relatively rigid structural and behavioural patterns' cope with their environment, he writes, 'determines which ones survive' (ibid.: 144). In another discussion, he expands on this question of organizational rigidity. Government organizations are seen as having 'an indisputable hardiness' (Kaufman 1976: 65). Rights, privileges and monopolies which government agencies often acquire, 'their protective armour', can 'turn into prisons' for them (ibid.: 12). Structured in such a rigid way, it can hardly be a surprise if even relatively youthful political systems, like those of America, find it far easier to deal with change by a process of incrementalism, adding to what exists and to what is allowed to persist, rather than by redrafting forms in a radical way (Winner 1975: 70–1).

Speaking generally, the responsiveness of organizations to change is affected by the extent to which their form and behaviour can be regarded as directed or steered. So far, I have treated organizations as having a common form. In reality, they can be arranged into an elaborate classification of types. A simple distinction is relevant to what I want to argue here. Some organizational forms, such as those embodied in patterns of administrative order or firms, have a degree of incorporation, coherence and co-ordination that provides them with a degree of steerage. In theory, their behaviour is capable of being responsive to change. Other types of organization, though, such as trading systems, need have no overall control or command function. Their political and social contexts might determine rules of operation, but not how they operate overall. Likewise, the behaviour of their respective parts might experience some conditionality, or some degree of interdependency, but there is no institutionalized form of steerage. Instead, their activity depends on countless individuals performing roles and routines in a loosely organized and co-ordinated way. Their co-ordination is as much informal as formal. Change for this sort of organization is not about the reprogramming of its strategic information by some centralized command or dominant elite, but about the simultaneous acceptance of a new form of co-ordination or new goals by its numerous parts or members. To recall a point made earlier, the asymmetry between the aggregate structure of the system and the part played by individual agents makes system-wide change in this sort of organization a more difficult and complex process, one full of potential lags. Yet, as writers like Arrow and Kaufman have shown, more formally co-ordinated organizations, those managed by a clearly defined and legitimized authority, do not necessarily perform any better as regards change.

Institutional and organizational inertia: a systematic view

We can bring these various points together into a systematic understanding of how organizations develop into a state of inertia. It supposes four broad stages. The first is a stage of creative adaptation. This covers the initial definition of roles and the formulation of covering rules and norms in response to a particular resource or opportunity. In addition, networks of social, economic or political interaction evolve. These networks become associated with routinized flows of information and energy.

A second stage sees the elaboration of organizational structure and behaviour. This is achieved through the varied processes of differentiation, specialization, incorporation, hierarchization, linearization and centralization, as organizations struggle to increase their AR:PR ratio and to abstract strategic decision-making to higher and higher levels. These generally lead to more coherence and to hyper-forms of integration focused around specific goals. In time, the elaboration of organizational structure means a loss of freedom, with the openness of initial choice slowly giving way to the relative closure of choice as existing structures box in the new choices that can be made. As Arrow especially has noted, the loss of freedom that stems from the elaboration of organizations is bound up with the fact that organizations represent an irreversible investment of human resources in specific roles and codes.

The third stage follows on from the elaboration of structure. The more elaborate and specialized its structure, the more an organization can suffer from the loss of redundancy or unused freedom. For a time, organizations respond to this decline in new opportunities through the rise of distributional coalitions that are more concerned with expanding their share of output than with expanding production and creating new opportunities. With this shift, organizations and their parts effectively move from being system-serving to being self-serving.

The fourth and final stage involves the onset of maladaptation. As a state, maladaptation has two dimensions. On the one hand, there is a time-specific dimension brought about by a failure of positive feedback and a more general failure to respond to changes or new opportunities in their wider social, economic, political or physical environments. The outcome is a tendency for organizations to preserve what – in their wider context – appears as a conservative, inertial form. On the other hand, there is their place-specific dimension. Most organizations are anchored to specific forms and layouts. This fixity of position is part of their potential for inertia. Just as the resources embedded in an organization are not easily recycled into other organizations so also does it follow that they are not easily transformed into a different geography.

The geographical inertia of organizational forms

Following on from this last point, I want to explore further the extent to which the inertial propensities of organizations have a spatial dimension. Most use space to help establish a stable structuring of their roles, function, and interactions or their data-handling capabilities. Indeed, competing with rival organizations over resources of space provides a great deal of their *raison d'être*. There are relatively few organizations that can be described as aspatial, whose functioning has no geographically meaningful component to it. For the many and varied forms that incorporate such a component, though, we need to grasp how it reduces their responsiveness to change. Altogether, we can identify a number of ways in which they acquire rigidity through the way they are arranged in space.

Organizations derive their structure from the strategic or hard-programmed information that defines their roles, relations and routines, what Unger called their 'task-defining information'. Such information can be contrasted with tactical or soft-programmed information, or that which is gathered, processed and exchanged as part of everyday routines, or what Unger called 'task-executing information'. When we consider how institutions and organizations maintain themselves, how they persist despite changes in personnel, the core of what is carried over as memory is strategic or task-defining information. As a structural template, we can expect this strategic information to be a particularly powerful source of inertia. Part of the reason why lies in how it defines the different ways in which organizations have appearances or forms of representation (e.g. administrative units, resource and property rights, information-gathering centres) and which are usually held as symbols of their being. Clearly, such appearances or representations are to do with the way structure is mapped in a durable way.

When we look closely at the way organizational forms use space, we find that it provides them with a powerful arena of choice. We can see this best by considering how they evolve. Most originate as a loosely structured flow or exchange of information. So long as they remain simple or diffuse, they can operate by having all members interact freely with all other members in a matrix-like fashion, a diffuseness that may be inefficient or sub-optimal but which gives them flexibility over how information should be handled and processed. However, as exchanges intensify, there are huge efficiency gains to be made by organizing the flow of information along defined channels and through decision-making centres that are increasingly ranked as the system takes on more co-ordination. For most political and economic organizations, this functional rationalization is invariably a geographical rationalization, a restructuring of how their operation is ordered in space through the greater specialization, linearization and hierarchization of its tasks, flows and decision-making.

History abounds with examples. During the early seventeenth century, when Amsterdam rose to prominence as the base-point or centre for the emergent world system, its growth involved two kinds of restructuring. First, it succeeded in fusing flows of trade from the Baltic, central Europe and the Atlantic into a single system or market of exchange. Second, it greatly energized the new system, dealing with an estimated range of over 800 different products. Amsterdam's achievement was clearly one of co-ordination, one that it underpinned with a range of new institutional innovations, regarding how trading ventures were funded, their shares traded and so on (Smith 1985: 985–1005). In co-ordinating so much information around one centre, it created certainty out of uncertainty and enlarged opportunities for trade. Being able to deal in so much information allowed for a more efficient pricing of goods, prices that were referred forward or backward to their price in Amsterdam. Like all major shifts in world trading patterns, the outcome was a reduction in transaction costs, a reduction that had multiplier effects on trade itself (North 1981: 31). For Chandler, comparable changes in organization underpinned the rise of American industry and its trade in the world economy over the late nineteenth and early twentieth centuries. Starting with the railroads in the 1840s and 50s, new management techniques were devised that enabled the railway companies to cope with the logistical problems of stock, track maintenance, traffic flows and timetabling across vast distances. The skills and techniques of co-ordination acquired were diffused into other industries as the American economy became driven by companies that had to deal with multi-plant and multi-product production systems and with marketing systems that embraced large areas. Better and better co-ordination meant greater speed of throughput and greater efficiency in the use of capital investments. Arguably, such success matured directly into Fordism, the mass production of standardized engineered goods using assembly-line techniques, and supported by an elaborate network of forward and backward linkages.

Chandler saw it as an achievement in geographical co-ordination, one driven by the central problem of how industrial organizations could co-ordinate themselves in space (Chandler 1988: 208–9, 401 and 408).

The development of more linearized, specialized and co-ordinated organizational forms, though, has its costs. From the point of view of inertia, the prime one is the expenditure of choice, a commitment of the organization's structure to a particular form and orientation within its operating environment and to a more selective way of gathering, processing and transmitting information within that environment. Further, each elaboration of structure deepens this expenditure of choice or flexibility. This applies to administrative systems no less than to trading systems, to the firm no less than to the organization written into a regional economy. As Arrow would argue, the expenditure of choice is an irreversible expenditure in the sense that the capital or energy given to it cannot be recovered. In any switch to qualitatively new forms of organization, the information channels, decision-making centres, codes, plant, land uses and so on have to be scrapped entirely or adapted at the expense of substantial conversion costs. Because of their strong geographical emplacement and orientation, most aspects of an organization's structure are simply not portable, convertible or divisible in this way without considerable costs and risks. Further, as each phase of elaboration acts as a further constraint on the range of subsequent choices available for an organization and its agents, there is a tendency for organizations ultimately to play Slobodkin and Rapoport's existential game, that is, to make decisions that enable them to go on playing the game or to go on functioning *in situ* by reinforcing the choices already made even though they may no longer be best- or good-practice (Slobodkin and Rapoport 1974: 196). The 'push of sequence' works to bias outcomes (Unger 1987c: 212), a bias which many studies of organizational sequences have noted. Once given coherence through co-ordination, real qualitative change to an existing organization becomes something that can only take place at a system or general level. Apart from massive devaluations of the capital invested in prior organizations, and I use the term capital here in the broad sense meant by Bourdieu, their only easy or capital-saving route for ongoing change is through incorporation, becoming part of a still more elaborate organization, the quantitative as opposed to the qualitative route to change (cf. Hallpike 1986: 237–52). In all forms of human organization, growth by integration or incorporation represents a common source of change, enabling new levels of decision-making to be added at a still higher level.

By the very fact that they exist as extended systems of relationship and information exchange, institutions and organizations have a situatedness, a position and orientation within a wider environment of activity and resources. Even seen simply as part of their appearances, this situatedness is a powerful source of inertia because of the way it can be used to negotiate personal boundaries and identities in their dealings with other rival organizations. However, for many key societal organizations, their situatedness is also part of the way they function. Indeed, if we consider the different ways in which the scalar growth of information flow and decision-making can stress an organization, the solutions devised to those problems that challenge it spatially are more likely to be built into its structure than solutions to those problems that challenge it temporally. Of course, there is a sense in which these problems of growth are time–space problems, being as much about how long it took for information to flow between parts of an organization as about how far it had to travel and about how it should be geographically centred. However, the fact remains that the way in which organizations have rationalized themselves under pressure of growth has usually been about how they have geographically rationalized themselves, as processes like linearization, specialization, centralization and hierarchization denote. In so far as such processes create structure at the expense of choice, we can regard the particular geography which they produce as an intrinsic part of an organization's drift to inertia. Seen in this way, landscape maps the geography of organizational constraint, its expended choice over resources and relationships.

7

The built environment as a source of inertia

Spatial fixity – the pinning down of locales to definite built environments, especially in the form of cities – marks a new departure in human history.

(Giddens 1984: 260)

Urbanization means the creation of relatively permanent resources systems. Human effort is, as it were, incorporated into the land as fixed and immobile capital assets that may last hundreds of years.   (Harvey 1985: 64)

From the moment societies began to adopt settled economies, their human landscapes acquired built forms. Whatever their function, all built forms represent a fixed store of capital, or energy, one that can only be recycled in a marginal way. For this reason, they amount to a powerful source of potential inertia. This potential, though, has varied over time and between cultures. We can capture some of its historical and inter-cultural variation by drawing a broad distinction between pre-modern and modern societies or between societies dominated by traditional forms of exchange based on reciprocal or regulated exchange and those fully penetrated by free markets.

The built environment and pre-modern societies

In a discussion of how pre-modern and modern towns differ, Vance has proposed that land in the former was perceived as having only a use value whereas in the latter it was seen as having an exchange value that overrode any use value (Vance 1971: 101–20). This distinction has a bearing on the inertial properties of the built environments produced by them. In a society in which land and the manifold ways in which it can be improved through investments of labour and fixed capital is considered in terms of its use rather than exchange value, then the potential life span of the built environment is its functional usefulness not its ability to yield a better than average rate of return on capital under a given set of market conditions. For the different forms of

139

the built environment, this functional usefulness covers their physical working life, not their profitability within a system of competitive markets. In other words, once a physical investment of capital was made and incorporated into the built environment, it continued to be used for the duration of its physical working life. This primary concern with the use value of its different physical forms gave the built environment of pre-modern societies a much greater capacity for persistence and, therefore, for inertia when compared with modern societies.

Of course, against this, it could be argued that we need to take into account the fact that pre-modern societies generally had fairly low levels of investment in the built environment, at least when compared with those typical of modern societies. This broad generalization, though, needs to be qualified. Though absolute levels of fixed capital can be seen as modest in relation to modern societies, they could still be substantial when seen as a proportion of all available capital. This was especially true of historical empires. By their very nature, empires extracted vast amounts of energy in the form of labour, food, raw materials, craft products and taxes from a wide area. Further, through their extensive use of *corvée* and slave labour, they could invest considerable amounts of labour in building projects, such as the irrigation schemes maintained by the hydraulic civilizations. Overall, they re-distributed wealth from peripheral to core areas (see, for example, Blaut 1992: 41–3). Invariably, their cores became major centres of consumption, converting part of this surplus into a fixed form, and enabling cities to be built that far exceeded the wealth of their immediate hinterland. Their capitals took on the form of imperial cities, reifying the wealth extracted from the empire as a whole (Sutcliffe 1983: 240).

Though political and military coercion was instrumental to this flow of wealth from peripheral to core areas, so too were the activities of merchants. Detailed analyses of mercantile wealth have revealed much about the general nature of wealth prior to large-scale industrialization. One of its distinguishing features was the extent to which the capital invested in business or trade was largely in the form of circulating rather than fixed capital. What is more, the kind of fixed capital involved (i.e. mills, warehouses, boats, wagons, materials, manufactured goods and the like) was flexible enough to be used in the context of different types of business activity. This is borne out not only by analyses of the capital tied up in particular business ventures, but equally, by analyses of individual patterns of wealth holding. In an analysis of English merchants of the seventeenth century, for instance, Grassby stressed its diversity of form, with wealth spread across a variety of different assets but with a high proportion capable of being liquified (i.e. gold, silver, jewellery) and a relatively modest proportion locked into fixed capital forms (Grassby 1970: 87–107). Such conclusions are echoed in Schama's analyses of wealth in seventeenth-century Holland, a period when it developed 'the most formidable capitalism the world had yet seen' (Schama 1987: 323). Though there was 'as much reason to describe it as a spend-and-prosper economy as a save-and-prosper economy' (ibid.: 322), he makes it clear that a great deal was spent on 'absolute consumption' rather than on durable assets, citing examples of the high level of wealth that could be spent on food and movables like jewellery and furniture. Yet that point made, he still concedes that some spent vast sums on housing themselves. The point that comes across from his discussion is that whilst the general character of capital in pre-modern economies was to consume it or to maintain it in a flexible, adaptable form, the supercharged levels of wealth associated with major centres of trade invariably led to some merchants solidifying some of this wealth into significant levels of fixed capital accumulation.

We can see this most dramatically in the context of Italian cities like Florence and Venice, both cities whose townscapes were greatly embellished by trading profits turned into the durable form of stone palazzi. Goldthwaite's study of Florentine merchants and their numerous palazzi stressed 'the fundamental fact that has to be understood why men wanted them is their monumentality' (Goldthwaite 1980: 102). They were built to be conspicuous and to last. Their monumentality though, 'was also an economic fact' (ibid.: 102). Indeed, Goldthwaite estimates that merchants were investing between one-third and one-half of their personal wealth in the building of palaces, and that, despite the belief that they should symbolize the persistence of the family, a significant number were sold within a generation due to the financial problems caused by the very demands of their construction (ibid.: 102; see also, Herlihy 1967: 6–7). In theory, we might expect such forms to be strongly inertial, since their residents would not be able to replace them, like with like, once stripped of the empire that funded them. Yet experience suggests otherwise. Many a Roman town was overwhelmed by less substantial later buildings. What must have been considerable capital investments, like the Roman amphitheatres at Arles and Nîmes in Provence, could be devalued as such by an infill of housing and rubble.

Taking a wider perspective, some have noted inter-regional or inter-cultural differences amongst pre-modern societies in their propensity for storing capital in a fixed form. One such difference has been highlighted by Jones. In many parts of Asia, he has argued, higher frequencies of natural disasters tended to inhibit the personal accumulation of wealth in the form of durable goods. By comparison, the relatively lower incidence of natural disasters capable of destroying property in parts of Europe meant that there was a greater marginal propensity for investing surplus wealth and energy in durable goods (Jones 1981). On this basis, we can expect a greater amount of capital or energy to be locked inertially into a fixed form in European than in Asian societies. We can qualify this conclusion in two ways. First, if we look closely at the frequency of natural disasters, they do not suggest a neat or tidy contrast between the two areas. Parts of the Mediterranean, for instance, were just as prone to destructive disasters as parts of Asia. Second, Jones himself has made it abundantly clear that whilst Europe was not afflicted by the same frequency or scale of natural disaster, the fixed and movable wealth of its towns were subjected to periodic destruction through man-made disasters, notably fire damage. Major fires ravaged the urban fabric of many European towns and cities over the medieval and early periods (Jones 1981: 33–4). Altogether, Jones, Porter and Turner listed over 518 known urban fires in England, 1500–1900, with most major towns being affected by extensive fire damage on more than one occasion (Jones, Porter and Turner 1984). Clearly, this would have militated against the role of fixed capital assets as a source of inertia though it does not follow that such fires automatically led to change. As with the Great Fire of London (1666), other potent forms of inertia, such as pre-existing property rights, were sufficiently entrenched to ensure a continuation of the existing ground plan, if not the buildings themselves, despite Sir Christopher Wren's plans for change.

Modern societies and the social construction of inertia

With the emergence of large-scale urbanization and industrialization from the late eighteenth century onwards, the levels of fixed capital sunk into the landscape rose dramatically both in absolute terms and in terms of the proportion of GNP embodied in it. Heavy investments in factories and their machinery, transportation systems, housing, public utilities like water, gas and electricity, civic buildings and so on, locked vast amounts of capital into particular configurations of geography and brought a whole new scale and dimension of inertia to landscape. However, there was also a change in the way inertia was constituted and defined by society. The root cause of this change was the maturation of free or price-fixing markets and their penetration into all sectors of the economy. The origin of price-fixing markets pre-dated the eighteenth century. A strong and convincing case has been made for seeing key sectors of English farm production as showing signs of price responsiveness as early as the thirteenth century (Campbell, Galloway, Keene and Murphy 1993). Detailed work on seventeenth-century London also suggests that there too, a bid-price market in land rents was in operation in densely occupied areas (Power 1986: 211–12). Yet despite these early indications of free markets starting to allocate land to those uses and users prepared to bid more, we must keep in mind the fact that other factors were still in place regulating and inhibiting market behaviour, especially as regards traditional crafts and trades (Dodgshon 1991: 261–4). What finally released the bonds of regulation and custom was the vast increase in new sectors of manufacturing and in other new spheres of investment from the late eighteenth century onwards. Only by having free access to land, labour and raw materials could industry secure the viability of its now vastly increased scale of investments. However, just as price-fixing markets create location by differentiating between sites and their accessibility, so also do they differentiate between old and new technologies, techniques and processes of production through their productivity or price competitiveness. Inertia, the failure to upgrade production functions continually in step with what markets determine to be average rates of return, now carried a more severe economic penalty. As free markets co-ordinated the use of new spaces and their physical improvement through various types of investment, the rigidity inherent in the fixed capital forms of older vintages became more and more of a handicap.

A tension developed between the growing amount of capital sunk into fixed inertial forms and the tendency for new investments and new innovations to shift production functions further and further to the right and, in doing so, to redefine what constituted an average rate of return on capital and what might be deemed viable under prevailing market conditions. The more capital was risked on heavier and heavier forms of fixed capital investments, and the more competitive markets enhanced the tendency for it to acquire obsolescence, then the more a conflict arose between ends and means, between the expectation that investment capital should continually seek out better than average returns on investment and the need to freeze it around specific forms of investment in order to exact such returns (Harvey 1982: 393–4). Clearly, the differential flexibility between circulating and fixed forms of capital created difficulties that threatened the whole system of capital movement and supply unless new ways of disinvesting were developed.

By way of a reply, it could be argued that the problem of declining returns on investment had always been a problem even under regulated markets, simply because obsolescence and a decline in relative performance or output had always been latent in whatever techniques or processes were used. The market conditions that began to emerge over the eighteenth and especially the nineteenth centuries, though, brought wholly new factors into the equation. The scale of production increased and so too did the size of area over which goods were marketed so that it became easier for efficient producers to displace inefficient ones. The efficiency of markets increased (North 1981: 33–44) and so too did the rate of innovations, both as regards product and as regards process, so that the speed at which capital forms became obsolete during cycles of innovation was sharply increased, leading to even greater fallout or redundancy in times during the intervening phases of depression (Schumpeter 1971: 37). In effect, inertia acquired a shorter and shorter time-fuse. Finally, but by no means least, changes in the sheer scale of investment in fixed capital forms served to raise the social cost of inertia, turning what had been a source of profit into a source of loss. Some would see the devaluation of capital forms that are still materially useful as amounting to a 'moral depreciation of capital' (Weeks 1981: 180).

When these various points are brought together, it need hardly surprise us that capital markets explored ways of preserving flexibility, of being able to create geography out of fixed capital and to profit from it, but not to share in the problems of its potential inertia and obsolescence. Their response was not an immediate one. After all, markets too had their inertial tendencies. However, they eventually responded in two ways.

Changes in capital flow

First, new ways of capitalizing projects were sought. As a lead sector for these changes, the British experience can be used to define how they unfolded. Prior to the nineteenth century, funding for any sort of trading or manufacturing enterprise was based largely on the principles of partnership. In shipping, merchants would club together to provide the venture capital for a particular voyage, the partnership being disbanded at the end of each venture. In key branches of overseas trade, such partnerships were superseded, though not replaced, over the late seventeenth century when major trading companies like the East India Company were established as ongoing partnerships, or joint-stock companies, and when dealings in their shares, along with government loans and securities, became the basis for the formation of the London Stock Exchange. The appearance of trading companies based on shareholding, with shareholders being able to invest or disinvest their capital, was an institutional innovation that had first developed during the seventeenth century in Antwerp and Amsterdam.

Significantly, the formation of joint-stock companies in Britain was an innovation that did not spread immediately into other sectors of the economy, except where the government licensed them and secured their position through a monopoly of trade such as with the Mineral and Battery Works (1568), a mining and smelting business. In fact, when the joint-stock company eventually did spread as a more flexible instrument of investment, it did so in the transport sector rather than in manufacturing. This might be expected given that in most modernization processes, the heavy capitalization of transportation systems tends to run ahead of large-scale industrialization. In Britain, the need to devise new ways of raising capital became evident with the drive to extend the road system over the eighteenth century. The solution adopted was the turnpike trust, which enabled capital for road construction and extension to be drawn directly from the local community and through tolls. As the demand for better transport spread to canals, the scale of capital needed rose substantially. Each new canal was constructed through a joint-stock company established for that purpose. During the canal mania of the 1790s especially, trading in the shares of these canal companies generated vigorous regional markets in capital (Ward 1974). By the 1830s, railways offered the prospect of a still faster and more flexible system of transport. Again, the heavy demands which it made on capital were provided by means of a joint-stock company set up by Act of Parliament. Whereas the canals had largely drawn on a regional market in capital, speculation in the railway companies opened up a national market in their shares, especially during the railway mania of the 1840s (Hawke and Reed 1969: 276). This was a critical moment, for it illustrated how once in place, the instrument of shareholding allowed an extended circuit of capital flow to emerge that preserved its mobility despite the huge amounts actually committed to lines, stock and buildings (cf. Kuznets 1955: 55).

Changes in the way industry was funded were phased in even later. The heavy and longstanding commitment of capital needed by industrial enterprises involved in activities such as mining and metal-working initially relied on small business partnerships like the Foley Partnership, a partnership that combined mines and blast furnaces in the Forest of Dean with a network of boats and wagons needed to transport pig iron up the Severn to the myriad furnaces and forges which they operated in the Black Country (Johnson 1951–2: 322–40). In fact, such partnerships survived as the prime means of funding industry until well into the nineteenth century even in lead sectors like cotton (Cottrell 1980: 19). The persistence of the partnership as an institutional form of both funding and management was significant for it meant that in most manufacturing ventures, capital and entrepreneurship were fused, with fresh injections of capital being provided by the re-cycling of profits (Chapman 1970: 248). Partners could not withdraw their capital without also devaluing it and without writing off the entrepreneurship which they had also invested. Admittedly, this did not prevent some of the more successful, like the Arkwrights, from shifting profits sideways into landed estates (Jones 1974: 160–83). However, no attempt was made to maintain exact costs or to disaggregate the different types of capital involved, such as fixed capital, whilst profit was actually viewed as a reward for entrepreneurship rather than for invested capital (Pollard 1965: 183 and 235).

In fact, it was not until the middle decades of the nineteenth century, when a shift to still more intensive forms of capital investment began to redefine the level of capital needed, that real changes were introduced. Following an Act of 1854, industry was allowed to draw in capital through the formation of limited joint-stock companies and the issuing of shares. The amount drawn in via the stock market, though, remained endemically low throughout the nineteenth century. Indeed, most of the industrial companies floated over the second half of the nineteenth and opening decades of the twentieth centuries were floated on provincial stock markets like Manchester and Liverpool where local perceptions and loyalties were instrumental in sustaining a flow of capital into industry (Edelstein 1971: 86). When we look at the far larger amounts of capital flowing through the London Stock Exchange during this period, the picture is quite different. Despite a further scalar jump in the investment needs of British industry as some sections of it, notably iron production in the 1850s and 60s and chemicals in the 1880s and 90s, pursued a policy of capital deepening (Cottrell 1980: 11), it was actually starved of capital by the Stock Exchange. In fact, an analysis of the pattern of business on the London Stock Market in 1914 suggests that not only did the total amount of trade in home mining, transport and industrial shares amount to only 8 per cent but in domestic industry alone, it was only 0.5 per cent of all its activity (Pollard 1985: 493). Most of its activity was in overseas shares, as capital flowed out to more lucrative green-field investments overseas (Ingham 1984: 146–7). Given how much profit had been derived from other activities in its overseas markets, it does not follow that this exodus of capital necessarily involved a large-scale migration of capital out of the increasingly obsolete and inertial forms of British industry through the now sharpening distinction between fixed and circulating capital (Usher 1955: 543). However, part of it can certainly be described in this way. As such, it represents an early and notable instance of capital's inclination to work a 'spatial fix' when returns on capital suffer a decline, to seek out greater returns overseas (Harvey 1981: 1–12).

Defining fixed capital: fixed in reality, elusive in concept

The second change in the nature of capital that needs to be noted is that, as the proportion of fixed capital was expanded over the nineteenth century, and as the risks of committing so much capital to a fixed form increased, changes took place in accountancy practice. Previously, few British businessmen had maintained reliable accounts or made any allowance for the depreciation of assets (Pollard 1965: 238). The jump in capital needs apparent by the mid-nineteenth century helped to change this (ibid.: 216). Businesses set out to cost investments more exactly, and an allowance was made for depreciation. Gradually, as the volume of capital rose over the second half of the nineteenth century, and as the amount of external capital brought in through the formation of limited companies rose, the practice of amortizing investments – depreciating fixed working capital at a set rate of cost to cover the fixed capital involved – became the norm. In the process, it became easier for the original investment to be valued and to be recouped with interest for re-investment. Though there were good reasons why industrialists both wanted and tried to lengthen the working life of existing stocks by improvements (Weeks 1981: 186), the return of investments plus interest back into the circuit of circulating capital invariably led to its switch into newer and better-yielding technologies elsewhere.

The problems which early industrialists faced over measuring the amount of fixed capital in their business and how precisely it should be depreciated, are highlighted by the fact that even amongst governments and economists today, no standard treatment for the analysis of fixed capital has found acceptance. The prime difficulty is that at any one time, fixed or commodity capital embraces buildings, equipment and other stock of varying vintages. Different commentators and governments have responded to this variety of vintage in different ways. In actual practice, we can find rates of fixed capital formation measured in terms of gross fixed capital, with the original value of each commodity or unit of fixed capital being assumed to remain in stock until retirement, or in terms of net fixed capital, with an allowance being made for its gradual depreciation (Kuznets 1955: 23; Griffin 1979: 102). In theory, though, a greater range of interpretations is possible. Rymes, for instance, has distinguished between analyses in which no depreciation takes place, with fixed capital stock assumed theoretically to remain 'in productive service forever and can, therefore, continuously and costlessly . . . be transferred from association with older techniques of lower output per unit of labour to newer techniques of higher output per unit of labour' (Rymes 1971: 61); depreciation 'by evaporation' or a 'constant force of mortality' (ibid.: 62); depreciation 'by sudden death' (ibid.: 63); and, finally, depreciation 'by obsolescence' with the value of fixed or commodity capital declining through an assumed ageing process that gradually reduces output to zero (ibid.: 66).

Seen from other perspectives, the treatment of fixed capital by economists can appear contrived. This is especially true when we are faced with measures of net fixed capital. Following the standard procedure used in firm accounting, measures of net fixed capital discount or amortize fixed capital investment at a fixed rate over what is deemed their useful life. In theory, this reduces their asset value to the firm to zero after so many years, and does so in a definable way. In practice, it is greatly complicated by the fact that different industries and different kinds of stock or machinery within them have widely different definitions of what this age expectancy amounts to (Griffin 1979: 99 and 114). That for all buildings involved in UK manufacturing, for instance, is fixed as eighty years, whilst that for machinery varies between twenty and thirty years (ibid.: 114 and 126). Clearly, this enables the complexities of fixed capital wastage to be factorized, given a value, but it also has the unfortunate side effect of assuming all elements of the built environment that have exhausted their expected lifespan – that is, a high proportion of such assets – to have no asset value to society. Of course, this would be an excessively crude interpretation of what the amortizement of investments is meant to achieve. It does not remove fixed capital investments but simply – in theory if not in actuality – reduces them to a 'costless' factor of production once their capital value has been depreciated, though there will still be maintenance and service costs (Rymes 1971: 66). Clearly, whilst such measures of fixed capital are relevant, they cannot easily value the extent to which old stock actually survives in use. The difficulty, as Hicks pointed out, is that most analyses of capital are time-set, or static – they have to deal with the capital that is available at each moment. They are not intended to lean backwards or forwards (Hicks 1973: 177). Of necessity, such analyses have to be blind to persistence or inheritance.

In reality, though, particular structures or processes may be very longstanding yet still be relevant to factors of production. Hicks himself refers to the
contribution to inputs still made by former Roman roads or the first clearing
of the woodland. 'If this investment,' he confessed, 'small though it may have
been by modern standards, is left to accumulate over centuries at compound
interest, it must dominate *capital at cost* in a way that is clearly preposterous'
(ibid.: 159). If we share his perspective as an economist, we must agree.
However, if our primary interest is in such hidden or neglected inertia, his
point takes on a different meaning.

Yet the practice of discounting fixed capital investment in this way does
make a vital point, one which some interpretations have seized upon. Fixed
capital is that part of capital that does not circulate through the circuit of
capital. As Weeks put it, it is 'that part of the means of production which has
a life longer than one circuit of capital' (Weeks 1981: 177). Whereas conventional economic theory would see this embeddedness being overcome through
a process of constant depreciation or discounting, some see its value being
released through crises during which fixed plant is dramatically devalued and
rendered socially useless (e.g. ibid.: 181; Harvey 1982: 393). Both approaches,
though, are making the point that the embeddedness of fixed capital, and the
potential inertia borne out of this embeddedness, creates problems, place-
specific problems, which industry has to overcome if it is to remain competitive and which capital markets have to overcome if they are to retain control
over the circulation of capital.

Fixed capital and landscape: a recent history

These problems of embedded capital can be more fully appreciated by considering how the importance of fixed capital has developed over time. Just as
Rome was not built in a day, so every society has had to live with a substantial
degree of inherited stock from the past. Locally, towns and industry may experience a sudden mushrooming of growth such as occurred in centres like
Middlesbrough and Barrow during the mid-nineteenth century or with new
towns in this century. Likewise, exceptional events, like the extensive destruction of European towns during the Second World War, may create a large-
scale need for the reconstruction of the built environment. Poland and parts
of eastern Germany, for example, experienced the destruction of almost two-
thirds of their urban fabric during the Second World War so that rapid and
extensive reconstruction after the war was essential. More usually, additions
made to the gross stock of the built environment during any one generation
or any one cycle of growth are marginal when seen in relation to the absolute
amount of fixed capital already embedded in the landscape. Yet notwithstanding the exceptional circumstances just noted, the extent to which this
inherited stock is carried over from one generation or one cycle of investment

to the next has increased, not decreased, historically. The very fact that absolute levels of net fixed capital formation have fallen since reaching a peak in the early decades of the twentieth century in most parts of Europe and North America is indicative of this growing inertia of fixed capital stock.

The relatively modest levels of fixed or durable capital that typified the pre-modern human landscape began to change with the onset of large-scale industrialization and urbanization from the late eighteenth century onwards. If we look at Britain's experience first, this upswing in the amount of fixed capital began slowly. Though the shift to mechanization and to factory production was underway in lead sectors like the cotton industry by the 1760s and 70s, the earliest factories were largely adapted from other uses, for example water mills, warehouses and even country mansions (Chapman 1970: 237 and 252), rather than purpose-built factories. When the first purpose-built factories were built over the 1780s and 90s, they raised the level of fixed capital but, as a proportion of the total invested capital, the amounts involved were still modest (Pollard 1964: 301–5). According to Chapman, the entire fixed capital of the cotton industry, a world leader at this stage, totalled only £2.5m in 1795, of which £1.25m was in the north-west of England (Chapman 1970: 247). For comparison, the largest geographical concentrations of personal wealth at this point were to be found in the trading centres (e.g. London, Bristol), not the manufacturing centres. The build-up of heavy investment in British industry did not commence until the 1830s and 40s, when there began a shift in the cotton industry to large, multi-storeyed, multi-process factories relying on steam- rather than water- or hand-powered machinery (see, for example, Cottrell 1980: 19). Other industries, like the extractive and processing industries, witnessed a similar upsurge of investment in durables by the 1830s. Major instances of capital deepening occurred later with the opening of the Cleveland iron ores in the 1850s and in the north-east chemical industry during the 1880s (Cottrell 1980: 11). It was at this stage that capital began to be drawn in through the formation of joint-stock companies.

Running ahead of these mid-nineteenth century surges in the build-up of fixed capital in industry was a dramatic growth of investment in new transportation systems. Indeed, in other countries, the construction of these new transportation systems consumed more capital than industry itself during the initial stages of industrialization (Cottrell 1980: 4; Feinstein 1981: 128). In pre-modern Europe, systems of land transport were sustained by the wagon, pack-horse or pedlar. So long as trade depended heavily on organic raw materials drawn from small scattered production points, such as farms, woodland cottages and town workshops, then transport systems needed to be equally diffuse and low cost. With the onset of industrialization and the growing use of organic raw materials from mines and quarries, the demands on transport shifted. What was now needed was the facility to move large quantities of low-cost materials between relatively few points as production became punctiform, concentrated around a limited number of centres (Wrigley 1962: 1–16). The response was the build-up of capital-intensive forms of transport: paved roads and turnpikes, canals and railways. The scale of capital consumed by canals and railways rivalled that being built up in the cotton and woollen industries, but to contemporaries, it appeared all the more dramatic because of the speed with which investment was switched into them. During the canal mania in the 1790s, vast amounts of local capital were 'sunk' into a greatly expanded canal network (Ward 1974: 73). The shift of capital into the railways during the railway mania of the 1840s was even more dramatic. Between 1841 and 1860, railways absorbed over 10 per cent of total GNP and 21 per cent of all fixed capital investment in Britain (Feinstein 1981: 133). As the absolute and relative amounts of investment in British railways subsided over the remainder of the nineteenth century, capital looked for new opportunities. The great flood of British capital overseas during the closing decade of the nineteenth and opening decades of the twentieth century helped to fund the construction of railways in countries like Argentina and India, again reifying huge sums into a fixed and, in capital terms, immovable form.

At the very same point at which so much capital became locked into the railway network of Britain, equally large sums were poured into the mould of a greatly expanded urban system. Significant sums had already been sunk into towns over the seventeenth and eighteenth centuries as many switched to brick and made other improvements of layout (Jones and Falkus 1979: 193–233). By the nineteenth century, their sheer physical expansion lifted the scale of capital involved still further. In fact, nineteenth-century investment in the urban parts of Britain's capital stock far exceeded that given to industry. We can identify two components to this investment. First, there was a vast flow of capital into house-building. As Whitehand pointed out in his analysis of house-building activity in cities like London and Glasgow, this flow was phased in step with the booms and slumps in the economy at large (Whitehand 1972: 39–55; 1975: 211–24; 1987: esp. chapters. 3, 5–6). In his stress on how the growth of housing moved outwards around cities in a series of annular rings whose phasing and scale was in step with the intensity of booms in the economy, his work underscores the marginal way in which major phases of investment add to the total stock of fixed capital. If we take measures of gross investment in housing, there are signs of a surge in the years immediately following 1800. After reaching a peak in the 1830s, flows of capital slowed down over the 1840s and 50s before rising sharply to new peaks in the mid-1870s then subsiding to a trough in the 1890s. By the turn of the century though, a renewed burst of activity is evident, one that surged forward until brought to an end by the First World War (Feinstein 1981: 128–42). Overall, when calculated at 1850 and 1914 prices respectively, the volume of stored capital in housing rose nearly threefold between 1800 and 1850 and again between 1850 and 1914 (ibid.).

The second component to the build-up of fixed capital in towns was that sunk into public utilities and facilities. The scale of urban change after 1800, especially after 1820, created the need for the more organized provision of basic utilities. The provision of safe drinking water and better sanitation was a longstanding problem that was now made all the more pressing by the over-rapid rate at which towns began to grow by the 1820s and 30s. Along with paved roads, the installation of piped water and water closet systems added substantially to the costs of urban development and the capital commitment built into the street plan of the typical town. The opening years of the nine-teenth century also saw the introduction of gas lighting. At first, its installa-tion was put in the hands of private companies, a strategy that meant the duplication of basic facilities as companies competed for the same customers. London, for example, had fourteen different gas companies by the middle of the century. By this point, though, towns had become more self-conscious and more concerned over how such basic utilities like gas lighting and water were being managed and delivered. As the second half of the nineteenth century progressed, the more assertive responded by taking over responsibility for the provision of basic utilities like gas, water and sanitation. This new role as pro-viders and managers of services was further expanded through a sequence of legislation that broadened the scope of their powers and responsibilities. Legislation empowered them to provide schools, hospitals, tramways (1870+) and electric lighting (1882+). As noted in chapter 4, towns added further util-ities that ranged from bathhouses to cemeteries. The sum effect was a colossal expansion in the amount of durable capital underpinning urban layouts, part sustained by municipal authorities themselves and part still in private hands (Falkus 1977: 134–56). Indeed, when we look at the total stock of fixed capital present in Britain by the eve of the First World War, that attributable to public utilities was one of the largest items in the account, with companies like the London Metropolitan Water Board being amongst the largest in Britain. Their accumulated investment helps gives support to Sutcliffe's point that Britain's advantageous exploitation of a world economy over the second half of the nineteenth century, like empires before it, encouraged a heavy physical investment in its urban fabric, an investment skewed towards the larger cities (Sutcliffe 1983: 240).

Using the figures compiled by Feinstein and Pollard, we can survey the trends of fixed capital formation in Britain from 1760 to 1920 (see Figs. 7.1–2). The overall levels of both gross and net fixed reproducible capital were rising gently by 1760 when Feinstein and Pollard start their series. The rate of increase steepens noticeably during the early–middle decades of the nine-teenth century. This high level of accumulation continues until the First World War. Only then are there the first real signs of the rate of increase slowing. Seen in relation to gross and net levels of national wealth, each increased at similar rates over the second half of the eighteenth and the first half of the

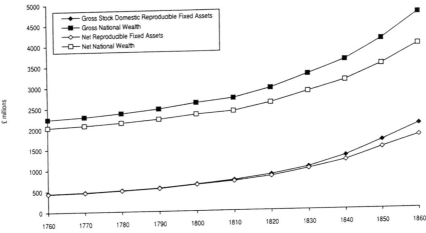

Fig. 7.1 Fixed reproducible assets and national wealth, Great Britain 1760–1860 (at constant prices, 1851–60). Based on Feinstein and Pollard 1989: 468–9.

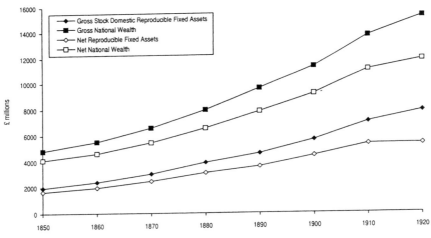

Fig. 7.2 Fixed reproducible assets and national wealth, United Kingdom 1850–1920 (at constant prices, 1900). Based on Feinstein and Pollard 1989: 468–9.

nineteenth century. By the 1850s, though, the rates at which both gross and net levels of national wealth accumulated began to pull away from the rates at which both gross and net levels of fixed reproducible wealth accumulated, the gap being especially wide in the years immediately prior to the First World War. When we break the increase in fixed reproducible capital down into its main components (see Figs. 7.3–4), we can see that the largest concentration of net fixed reproducible wealth lies in dwellings and that its rate of increase

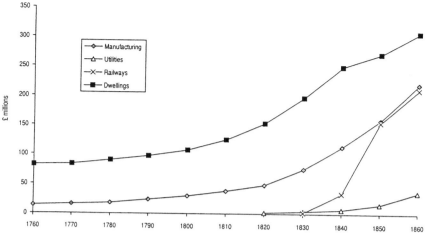

Fig. 7.3 Net stock of fixed reproducible assets by sector 1760–1860 in Great Britain (at constant prices, 1851–60). Based on Feinstein and Pollard 1989: 452–3.

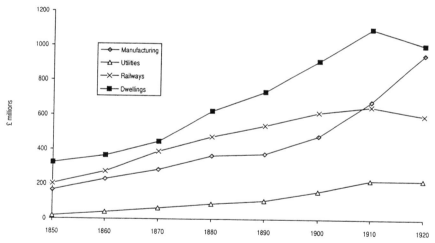

Fig. 7.4 Net stock of fixed reproducible assets by sector 1850–1920 in the United Kingdom (at constant prices, 1900). Based on Feinstein and Pollard 1989: 452–3.

rose gently in the second half of the eighteenth and more steeply in the nineteenth century, especially once urbanization gathered momentum from the 1820s onwards. The amount of net fixed capital locked into industrial uses was always less than that in dwellings, but its rate of increase displays the same sort of profile. The main difference lies in the way the formation of fixed capital in industry began to rise more steeply by the 1840s and 50s. As Figs. 7.3–4 show, easily the most striking change in fixed capital formation over the period

1760–1920 was the dramatic increase in the amount of capital sunk into railways over the 1840s and 50s. For comparison, when we move forward into the twentieth century, the outstanding feature of GNP spending has been the greater and greater amount devoted to public services (Fry 1979: 2).

These changes in the trend and composition of British fixed capital investment over the eighteenth–twentieth centuries are matched by those of other countries as they too experienced the take-off of large-scale industrialization and urbanization. At a time when overall levels of capital formation increased dramatically, the percentage represented by stocks of fixed or durable capital rose substantially. In effect, the Industrial Revolution raised the level at which society everywhere consumed capital as a largely irreversible expenditure of resources on fixed structures of landscape. However, as with Britain, available data suggest that the proportion of GNP committed to fixed capital formation had levelled off by the end of the nineteenth century or the start of the twentieth.

In France, absolute levels of gross domestic capital formation covering investment in buildings, machinery and equipment and transportation rose sixfold between 1820, when economic growth began to take off, and the start of the First World War (Lévy-Leboyer 1978: 292–5). This was during a period when total GNP in France rose threefold (ibid.: 239). Seen as a percentage of GNP, the amount committed to fixed investment rose from just over 7 per cent between 1810 and 1819 to 15 per cent by the start of the First World War, though most of the increase took place in the decades down to 1880 rather than after (ibid.: 239) or during the decades when the shift into factory production and investment in railways was in full swing. Patterns of investment in Germany followed a similar path, with an early surge of investment in the durable stock of industry and new transportation systems more suited to the needs of an industrialized economy. Figures for Prussia show that in the periods 1806–22 and 1840–9, investment in industry rose almost threefold, that in transport over twofold, and construction also over twofold. In absolute terms, though, those devoted to buildings of any sort and to transport were by far the largest components of investment (Tilly 1978: 427). Figures published by Tilly enable us to see this not simply in terms of site improvement but in terms of expansion, with the number of factories increasing almost fourfold, that of public buildings doubling and private houses increasing by 50 per cent between 1816 and 1867, the initial stages of change (ibid.: 435). Figures published by Hoffman bring the pattern forward. Overall the rate of fixed capital accumulation in Germany peaked in the years immediately around 1900 before falling back significantly during the inter-war years, but then surging forward again in the years following the Second World War (Hoffman 1961: 124 and 131). When we examine its changing composition, we find that investment in residential construction was subject to strong cyclical swings, but overall it rose with the onset of large-scale urbanization to around

30 per cent of all investment in fixed capital, remaining steady thereafter. That invested in factories rose to around 20 per cent of all such investment. That directed at industrial plant and equipment rose to similar levels by *c*. 1900 and, after maintaining this level during the inter-war period, then rose further in the years immediately following the Second World War as reconstruction got under way (ibid.: 126).

Patterns of investment in America between 1869 and 1948 have been summarized by Kuznets. As a percentage of GNP, domestic gross capital formation hovered around 20 per cent per annum down to the 1930s, when it fell back to 14.8 per cent before an upward surge to 28.5 per cent during the 1940s. If seen in terms of domestic net capital formation, with allowances for depreciation built in, the percentage is around 13–14 per cent in the decades down to *c*. 1900 then falls gently over the next thirty years and sharply in the 1930s (2.3 per cent) before climbing back to 11.4 per cent in the 1940s (Kuznets 1955: 62). Despite these late surges, which were largely a by-product of the war effort, the verdict of Kuznets is that for every category, durable reproducible wealth declined at an increasing rate (Kuznets 1946: 59).

Taking a wider view of these trends in a sample of countries (USA, Canada, UK, France, Sweden and Denmark), he argued that increases in the rate of gross capital formation tended to occur shortly after, rather than coincident with, the initial increase in general rates of economic growth and then to decline slowly thereafter as the provision of social services and the greater personal consumption of perishable goods took a larger slice of GNP. This decline was apparent in each of these countries, Sweden excepted, by the years following the First World War. Only in exceptional circumstances, such as following the extensive destruction of fixed capital forms in Germany and parts of eastern Europe during the Second World War, has this tendency for rates of gross fixed capital formation in the older industrial countries to decline over this century been reversed (see, for example, Hoffman 1961: 124). Changes in the rate of net capital formation also display some variation. Kuznets's analysis shows that it also rose to a peak during the late nineteenth or early twentieth century then declined. He makes the point that over this century, depreciation of stock has become more important in countries like the USA and Canada when compared to a country like Britain (Kuznets 1955: 32). In all probability, this reflects endemic problems of the British economy, with the lack of investment by the late nineteenth century fostering a culture in which depreciation was set on a slower time fuse so that older stock persisted for longer.

Bringing this data on capital formation together, a first and obvious conclusion to be reached is that the emergence of large-scale industrialization and urbanization has worked to raise the degree of capital embedded in the construction of the human landscape. To this extent, it has worked to increase greatly the potential inertia of landscape and its built forms. The sharp increases in the proportion of GNP consumed as fixed or durable capital during the early stages of the Industrial Revolution, though, eventually gave way to a declining proportion as more and more capital was expended on government services and consumables. Even allowing for the fact that what has declined is the rate of increase not the absolute amounts of fixed capital locked up in landscape, this declining rate of replacement means that the latent inertial properties of such landscapes are slowly becoming actual as the built environment is replaced at a slower rate. We can break the overall increase in the rate of fixed capital formation down into different phases. The sharp increase in the rate of fixed capital formation that we can associate with industrialization and urbanization did not occur during the early stages of take-off. Initially, economic growth was sustained using existing techniques and plant, as Wrigley's stress on the continuing role of hand- and water-power in the early stages of British economic growth illustrates. In fact, it was not until well into the nineteenth century that a broadly based coal-fired economy developed, with all its implications for fixed capital investment (Wrigley 1988). This is also borne out by the fact that when the rate of fixed capital formation did increase, its early stages were marked by heavy investment in buildings (including domestic housing), transport systems, notably railways and dock systems, and other forms of public utilities like sewage and water supplies. As one writer put it, what distinguished early forms of fixed capital investment in the modernization process was their 'specificity' (Lowe 1955: 593). As a commitment of capital and resource, they were largely irreversible. Not surprisingly, some of the more intensive and specific forms of nineteenth-century capital formation, such as railways, have turned out to be amongst the most inertial. This is given some support by the sharp collapse in the capital coefficients of railways over the late nineteenth and early twentieth centuries, as the initially high levels of investment were inherited as a 'costless' factor into later accounts (Domar 1961: 113). It was only in the later stages that significant amounts of such investment began to comprise machinery. The fact that substantial portions of this machinery were flexible, thanks to the growing standardization of parts (ibid.: 95–117), meant that the problem of inertia was lessened proportionately.

Looking at the problem through capital coefficients, it has long been argued in regard to early capital investment that, initially, it concentrated on factor substitution rather than increases in output so that high coefficients were the order of the day, whereas later, as investment spread to more productive machinery, it also tended to raise outputs (ibid.: 106). Yet even for industries which gained considerably from new and more productive machinery, the build-up of capital investment in particular technologies could be carried forward on a powerful wave of inertia. This inertia is played out in different ways. Investment itself can be inertial, directed along flows that have ceased to offer the best returns. Gertler, for instance, has observed that whilst some see capital as shifting easily between opportunities, there are others who see its build-up in particular locations as cumulative with 'a scenario in which the spatial distribution of capital stock changes rather slowly over time and primarily in the direction of change established in earlier periods' (Gertler 1984: 151). Gertler's own view inclines to the latter, with 'inherited' stock having a negative feedback or reinforcement effect on investment (ibid.: 152; Walker 1985: 255–6; Storper and Scott 1989b: 27). Given time, regions will be abandoned by capital, as it seeks out opportunities to invest in wholly new technologies and products, or simply to tap a new labour force (Harvey 1982: 393; Walker 1985: 255). Even after fresh investment has ceased, though, existing industrial complexes can show a dogged persistence, working out the use-value of stock at below average rates of return.

Driving the migration of capital is what some have called the contradiction created by a struggle between different types of capital. In an effort to reduce the cost of production, industrialists may seek innovations that lengthen the usefulness of their capital stock, so that there will be a tendency for existing production systems to persist albeit at average or less than average rates of return. However, working against such tinkering is the effect of competition, with other industrialists trying to rejuvenate production functions by seeking out wholly new solutions and possibilities, ones which effectively devalue the usefulness of existing investment (Weeks 1981: 186). In the long term, capital has coped with these contradictions by seeking flexibility. At an investment level, the joint-stock company, a device designed to enable capital to be moved in and out of a business, became widespread. If the overall balance of such movement is towards disinvestment, then it could of course greatly weaken the market competitiveness of a business, though some have actually argued for a creative as well as a destructive side to disinvestment owing to reinvestment elsewhere (Isard 1977: 13). At a production level, companies have also sought solutions by working for more standardization of parts and equipment so that such capital becomes less fixed, more easily adapted to new opportunities (Lowe 1955: 583). Such standardization was already a feature of American industry by *c.* 1900. However, by this point, the emergence of large-scale assembly-line production systems, or Fordism, had started to redefine the scale of industrial fixed capital.

We can add a further dimension to this debate by using a point made by Adam Smith. Looking forward, Smith argued that as industrialization and economic development progressed, the rate of return from industry would decline as new technologies spread and rates fell back to an equilibrium level, but that the improvement of land would lead to an increase in rent and, therefore, in the proportion of national wealth accumulated by the landlords. Prompted by Lefebvre's version of this idea (Harvey 1985: 62), Harvey has suggested that during cycles of investment and disinvestment, or booms and slumps, there is a tendency for capital to shift between a primary circuit of investment in production and a secondary circuit of investment in urban forms, and that overall, the latter circuit and its culture of rent-seeking will become progressively more significant (ibid.: 62–89). That urban building progresses in cycles is not is dispute, but as Adams (1970: 37–63) and Whitehand (1975: 211–24) especially have shown, such cycles occur during economic booms, so that to a degree, speculative building in the housing sector competes with industrial capital. However, as Harvey explained, the growth of towns not only sunk huge amounts of capital into a highly localized and immobile form, creating additional Ricardian rent through improvement, but also created scarcity where none existed before by creating towns and cities ordinated around points of best and least-favourable locations. In doing so, it greatly multiplied the opportunities for extracting locational rent (Harvey 1985: 64). There can be no doubt that Harvey's basic point is both important and valid. If we look at urban land rents in a densely settled country like Britain over the main period of urbanization, they multiplied quickly. From £3m in 1845, for instance, the gross urban land rent of England and Wales rose to £49.2m by 1910. Whilst, as a percentage of national income, this involved an increase from 1.3 per cent in 1845 to a peak of 3.0 per cent in 1888, it then actually fell back to only 2.6 per cent (Singer 1941: 224; see also, Sutcliffe 1983: 258). Such was the Ricardian impact of improvement and location on rent by *c.* 1900, that an acre of built-up land in the city of London was worth 4,000 times more in rent than an acre of agricultural land (Offer 1981: 255). Figures for American residential real estate values and rents, 1889–1948, show that equally progressive and dramatic increases were also a feature there (Domar 1961: 112).

Yet we need to scrutinize the data more closely. If we look at relevant data on the rate of fixed capital formation from 1869 to 1948, it suggests that whilst the proportion of national capital formation provided by manufacturing rose sharply over the closing decades of the nineteenth and opening decades of the twentieth century, it fell sharply as the century wore on. The proportion represented by residential construction actually declined throughout the period under review, but at a lesser rate than for manufacturing. Indeed, from starting out at a level substantially below that of manufacturing during the early–middle stages of industrialization, it finished the period in 1948 contributing a greater proportion to national capital formation than manufacturing, 17 per cent as opposed to 12.5 per cent (Kuznets 1955: 56). The greater resilience of urban rents and construction can be attributed to land improvements and the increasing scope for surcharging rents with a bid-price for location. However, there is another aspect to be considered. Because of their large residential and civic component, urban land rents were less affected by rapid rates of obsolescence compared with manufacturing. Obviously, if the local economy did suffer from old, unproductive technologies so that it became less competitive and less attractive for investment, this would affect people's ability to bid for housing and land for other uses. However, its effect on rent was different if only because it was less direct. Initially, at least, so much land rent involved local circuits of capital, that there was less likelihood of capital being disinvested and redirected into new areas and spaces elsewhere. This is probably the reason why, in his analysis of land rents in England and Wales over the early and middle phases of urbanization, Singer found that their progressive advance was due more to their failure to adjust downwards in slumps than a tendency to rise rapidly in booms (Singer 1941: 229). In more recent decades, of course, this has changed, with urban property and rents becoming more and more a subject of investment by national and international circuits of capital. As this has taken place, the checkerboard of urban space has been opened up to sudden and dislocating shifts of capital no less than manufacturing.

To summarize, the process of industrialization and urbanization was accompanied by a substantial increase in fixed capital formation. A large proportion of this investment, especially during the early and middle stages, was locked into fixed and immobile structures. In other words, it was an investment that fostered inflexibility and inertia. Though different sectors – manufacturing, transportation, public utilities and housing – experienced this inflexibility to differing degrees, all were in some ways affected by it. As a problem, the persistence of built forms did not in itself affect their usefulness. So long as they went on working, they had a use value. The real problem was the capital locked into them. So long as trade and industry used re-cycled profits, or capital provided by the entrepreneurs who actually organized and ran them, then the idea of the capital invested in plant and equipment being disembodied or liquified like that invested in raw materials at the end of each cycle of account did not arise. However, as the scale of capital solidified in fixed capital forms rose, especially in sectors whose production functions were progressively transformed by innovation, it created pressure for change. The direction of this change had already been signalled by seventeenth-century trading ventures and the introduction of the joint-stock company based on shareholding, a device that allowed disinvestment as well as investment. When new transport and manufacturing technologies, together with the public utilities needed by a growing urban system, began to demand a new scale of investment over the nineteenth century, the principle of the joint-stock company became the logical instrument for drawing in such capital. Combined with the use of accountancy practices that debited an interest payment for such capital and the payment of dividends, investment capital could be reconstituted with interest. In the process, there now emerged a flow of circulating capital that could engage profitable activity without being penalized by the rigidity and inertia that went with fixed capital forms.

The effect was that whilst circulating capital could preserve its youthfulness by seeking out new and higher-yielding returns, the forms created by previous rounds of investment were abandoned to stagnate, starved of new injections of capital. In a highly competitive and market-driven economy, inertia was transformed into a burden of neglect. When we look at analyses of older areas of industrialization and urbanization, the different phases of investment and periodic shifts of capital, creating then abandoning place, are the dominant geographical themes. Over time, they manufactured a multi-layered landscape, with different layers of investment and capital functioning in each locality or region, each layer representing a different stage and geographical position within the investment cycle (Massey 1984: 118). In countries like Britain, there existed above these flows of provincial or national capital another layer that cherry-picked investment opportunities at an international level. With an acute awareness of such opportunities established through its empire, this layer of international capital was a major channel of capital investment for Britain.

Sectors like manufacturing continued to be afflicted by this endemic hardening in the arteries of capital flow. When new production strategies based on assembly-line production were adopted over the early decades of the twentieth century, the standardization of parts and machinery was seen as conferring some flexibility on the investments involved. However, there is a strong case for arguing that the degree of inertia attached to these new Fordist techniques of mass production actually increased owing to their complex organizational forms and the way they co-ordinated a complex array of inputs and outputs across time and space (Storper and Scott 1989b: 27). By the 1970s, new solutions were being experimented with in the form of so-called flexible accumulation, a system whose name strikes at the very heart of what, for free market capitalism, has always been the downside of making profit out of making goods, that is, making geography. Once any sort of system is established, a point is reached when further *in situ* investments are 'limited by the historically produced physical, social, political and ideological rigidities of the local milieu' (Swyngedouw 1989: 34). Sooner or later, the open-door strategy of moving capital to new centres and new technologies becomes the preferred alternative. As with earlier systems, flexible accumulation is not simply a new system, one which allows for a flexible switching of labour and output between different products and processes quickly and for levels of output to be raised or contracted in step with market demands with equal immediacy or which integrates or agglomerates an elaborate network of small specialized producers into 'a transactions-intensive production complex' (Scott 1992: 222). It also involves new spaces developing 'in geographical contexts that are insulated from older foci of Fordist mass production' (Storper and Scott 1989b: 27) and showing 'a general aversion to locations too close to the old centres of mass production' as part of its strategy for evolving new systems of production and labour control (Scott 1988: 106; Scott 1992: 222; for counter views,

see, Florida 1996: 315; Markusen 1996: 293–334). Though some have questioned the sufficiency of flexible accumulation as a concept, arguing that there
is more to the relocation processes involved than simply cost structures, few
would deny that it does highlight the problems posed by the rootedness of
modern manufacturing systems and their associated labour markets (Peck
1992: 334–5).

   Yet whether we are dealing with Fordism or flexible accumulation, each has
its own sources of inertia, or what some would now call lock-in or path dependency (Storper 1992: 78–90). Despite its name, flexible accumulation does not
really escape this dilemma. As Gertler put it, the 'price of acquiring production systems which are physically flexible is the adoption of inter-firm relations
which are far more rigid than before' (Gertler 1988: 425). The paradox is that
whilst changes in the cultural basis of society have led to inertial landscapes
because they are more valued, the working out of free markets has created a
counterclaim for seeing such persistence as a dysfunction. Yet arguably, whilst
free market capitalism treats inertia as a dysfunction, it gains enormously
from the disequilibrium or differences which this introduces into the geographical equation of growth. As Hirschman once said in the days before
restructuring dramatically raised the evident social cost of capital mobility, if
economic change is to be kept moving then the challenge of development
policy is to 'keep alive rather than eliminate the disequilibria of which profits
and losses are symptoms in a competitive economy' (Hirschman 1955: 66).

8

Conceptualizing inertia: the first step towards a geography of societal change

Culture is the organization of things in motion, a process of energy transformation.
(White 1959: 38)

. . . the hot society records itself in an essential way, on the world outside – on nature, on stone, on wax, on clay, on paper, on film, on tape, in its railway networks, its streets, its freeways – whereas the cool society is more nearly WRITTEN IN ITSELF
(Wilden 1972: 408)

In essence, inertia is about persistence, the exclusion or absence of change. Approaching the problem of change through inertia, therefore, may seem paradoxical. Treating society's tendency towards inertial states as a congenital condition hardly makes this paradox easier to resolve. Yet I want to argue that it is important that we recognize and define this paradox if we are to establish the foundations for a geographical understanding of societal change. By approaching the problem through the inertial tendencies of societal systems, we are made to appreciate how the analysis of change is as much about what constrains it as about the forces and pressures that are working to bring it about. It is not simply a one-sided calculation of benefits or advantages, as if there were gains but no costs. If it were, then change would appear as a seamless and frictionless shift into new forms and possibilities. Further, it would be largely ontogenetic in character, producing new forms from within itself. Given the historical and recent record of societal change, such a view is untenable. It is of the very essence that change involves an interaction between past and present, not just a wholly forward-looking, cost-free exploitation of present opportunities and their future potentialities. We need to construct interpretations of change that treat prior forms as both informing and constraining choice so that, at any one point, change becomes an interaction, a trade off, between what needs or wants to persist and what is open to change.

Yet when stated in such simple terms, inertia adds little to the debate that is new. However, if we add the extent to which inertia is not just a fundamental attribute of society and its forms but a geographically defined attribute, then the question of how change relates to established or inertial forms is not simply a matter of how it confronts and, on occasion, transforms them. If inertia is a geographically emplaced attribute of society and its basic institutional forms, and if the degree of inertia itself varies geographically, then we need to consider the possibility that societal change uses space strategically to circumvent areas or centres of greatest inertia and only as a last resort does it start by actually confronting these areas or centres of greatest inertia. Seen in this way, the geography of inertia becomes implicated in the twin questions of why and how does change take place, enabling us to reduce them to the single generic question of where does it take place.

Taking this argument further, I want to break it down into three broad interconnected themes. First, I want to consider inertia not as a dysfunction of society, as if it were something left as an abandoned and anachronistic relic from the past, but as a functioning and necessary part of societal systems and how they are constituted. This will prepare the way for the second theme which is a consideration of whether we can take the various sources of inertia and transform them into a single and internally consistent interpretation of their meaning. In a third and concluding theme to this chapter, I want to consider how inertia operates as a constraint on change and how this aspect might also be conceptualized.

Why is society inertial?

The treatment of inertia has always had an ambivalent role in debates over change. The fact that societies can stand still and can even employ change-avoidance strategies is recognized, but nevertheless, the prevailing view is that the analysis of change is seen as taking over where the acknowledgement of inertia leaves off. Inertia is consciously excluded from the debate as if it were the antinomy of change, part of the problem but having no part in the solution. As a result, no attempt is made either to define it or to factorize it in relation to change other than through the gross distinction which some have drawn between hot and cold societies, or those that change and those that do not. Despite the fact that different forms of inertia abound in the constitution of present-day no less than past societies, nevertheless, its interpretative significance is seen as becoming progressively diminished as societies seemingly cope with the challenges of modernity.

We can see this best in recent attempts to construct a society entirely of the moment. Thus, those who take a structurationist approach, with structure being reproduced recursively through everyday social practice, are really giving support to such a view because they effectively remove any actual distinction between what is open to change and what is not, or between what is inertial and what is of the moment. What matters to such approaches is that social structure, no less than what is process, exists only through each instanciation, or what is continually reproduced recursively through social practice. In other words, societal systems are seen as having no inertial properties in any meaningful sense because they are reproduced afresh through each instantiation. The way in which such a view can affect our handling of how the past intrudes into the present is put bluntly by Hindness and Hirst, even though one can hardly describe them as writing from the standpoint of structuration theory. After arguing strongly for a process of social reproduction, they note that the 'historical analysis of the "current situation" is impossible' because the conditions of existence for particular social relations 'are constantly reproduced in the present' (Hindness and Hirst 1975: 312). This is a remarkably ahistorical concept of the present, one which allows for no inertia, lags or relaxation time in the switch between qualitatively different structures or systems. Instead, it implies a world of absolute contemporaneity, with perfect equilibrium between prevailing processes and their social effects.

More aware of the materiality of social life, some geographical readings of structuration have tried to argue how each instantiation is historically contingent, bound up with prior conditions and the pathways by which society and spaces came to be. Such a view, for instance, lies behind Pred's particular handling of structuration theory when he talks about the historically contingent becoming of place and about the generative rules and power relations that constitute social structure being 'built into a specific historical and human geographical situation' (Pred 1984: 281; 1985: 337–8). Yet whilst there is much in what Pred argues that helps make my own case, his use of structuration theory does not entirely succeed in marrying past and present. Part of the problem lies in his stress on contingency. Strictly defined, contingency stresses the breaks, conjunctions and unpredictability of societal process or structure when viewed through time, not what might persist or be carried over. In so doing, it compresses the past, reducing it to something that is immediately antecedent rather than something whose depth has many layers and many exposures. To put this another way, we cannot have a notion of contingency that tries to be too historical, or deep-rooted, without weakening the notion of contingency itself. As Curry's analysis of the role played by chance concluded, if the development of landscape is 'a series of contingent events then in a certain sense the study of the historical record cannot reveal many of the generalities of process' (Curry 1972: 615; see also, Curry 1964: 146; Walker 1985: 227). Perhaps more serious is that Pred's use of structuration theory does not easily encompass the past in the way that he would like it to do. On the one hand, he talks about the 'sedimentation' of past forms, both those of the built environment as well as those of social and cultural practice (Pred 1985: 337; 1984: 284–5), but on the other, he sees 'outer nature, or the physical environment' as 'perpetually transformed' through the structuration process (ibid.: 243). Arguably, if we accept that any sort of 'sedimentation' or persistence is involved, then we are, in effect, qualifying the value of a structurationist approach, defining areas of life and landscape which are not 'reproduced' through social practice actually or metaphorically. Given that Pred's notion of historical contingency relies on these sedimented forms and processes, the case for exploring alternative ways of conceptualizing such 'sedimentation' and how it links in with present circumstances becomes persuasive.

Recent thinking by post-modernists like Bhabha and Jameson has taken this sort of argument a step further. For Bhabha, cultural meaning or signification exists in performative time, that is, in the moment of enunciation, so that – and here he quotes Fanon – 'it is always contemporaneous with its enunciation' (Bhabha 1994: 152 and 228). When he talks about how 'the disruptive temporality of enunciation displaces the narrative of the Western nation' (ibid.: 37), he is quite deliberately arguing that each new place of utterance or enunciation involves a 'disjunctive present' (ibid.: 254), a 'temporal caesura' (ibid.: 242). Jameson's reading of globalization would leave us in a similar condition. The nature of globalization, with its interpenetration of different spaces and temporalities, has, like Bhabha's reading of how the nation-state attempts to forge unity out of diversity, produced a break in the signifying chain and has brought about a contemporary failure 'to unify the past, the present and future of our own biographical experience or psychic life' (Jameson 1991: 27). Living amidst a 'rubble' of 'unrelated signifiers', we are now doomed to living our lives out as 'a series of pure and unrelated presents in time' (ibid.: 26–7). In fact, the rubble is imaginary, for in reality, 'the survival, the residue, the holdover, the archaic, has finally been swept away without a trace' (ibid.: 309). In short, for Bhabha and Jameson, post-modernist society is forever marooned in the present.

In reply, I want to argue that we can best begin to conceptualize inertia and the notion of an ongoing past by establishing the essential role which it plays in the normal, everyday functioning of society. In order to do so, we need to look afresh at how past and present are linked. It is a relationship that varies with scale. Many recent analyses have tended to be orientated around the individual's experience of the problem. As Urry has put it, when we reduce our perspective from an aggregate or systemic level down to that of the individual, a reductionist drift which lies at the heart of much recent social theory and a growing amount of theory in human geography, we introduce a greater tenseness into our perspective and an acute consciousness of what is past, immediate or future (Urry 1985: 23). A similar ego-based orientation of perspective is implicit in Giddens's imaginative use of presence and absence 'in the fading away of time and the shading off of space' (Giddens 1984: 283). There is no pure space, or *spatium*, to use Shields's term, only relational space, that is, a space that is 'fine-grained, myth-laden, qualitatively inflected, and *people-*

filled spatialization' (Shields 1992: 189). For Shields and others, though, the neat ordering of relational space around axes of near/far, inclusion/exclusion, presence/absence, has been undermined by globalization, with its 'collapse of the far-off into the nearby' (ibid.: 195; see also, Appadurai 1990: 295–310).

Such concepts enrich our grasp of the time–space ordination that surrounds the individual, but their meaning is altered when they are applied at the aggregate or societal level. This is especially so in regard to the experience of time. Seen by the individual or from the vantage point of a particular generation at a particular point in time, the course of history appears as a converging and increasingly predictable series, a chain of events, processes and possibilities that appear to narrow down to those which are manifest in the present, just as the future appears as a diverging series, with a broadening range of possibilities and uncertainty. Even worse, informed by aftermindedness, there is a tendency to squeeze all past trends into the template of the present in a teleological way, making its form as well as its moment in time appear inevitable. To adapt Shields's phrase, we can typify this as 'people-filled' temporization. As with an individual's genealogy, we are emphasizing a particular set of relationships out of a more complex network of ongoing relationships, a network whose wider character does not actually narrow or broaden in this way. At an aggregate or societal level, the richness and, more to the point, the openness of trends and possibilities is ongoing. It is only their orientation around individuals or generations that makes them appear otherwise. The increasing certainty with which we read the emergence of the present, though, has a further consequence. It fosters the notion that because what we see has been brought to a point of sharpest focus in the present, then it can be regarded as synchronous not only because it exists together within the present, shares a contemporaneity, but also, because it appears to have emerged as if their cohesiveness was planned or intended.

This leads me to what I want to call the synchronic illusion. The greatest problem facing the study of how the past intrudes into the present is the fact that we tend, whether as particular individuals or generations, to experience the past in a way that treats it as if it were synchronic when, in fact, it is richly diachronic. What survives into the present appears to each individual or generation as a given or functioning system, one which, to all intents and purposes, is of the present and, therefore, accessible to it. Part of the reason for this lies in the tenseness, the here and now, of our perspective. Part also lies in a point that was made by the philosopher Whitehead. Whatever the origin of what we see, it is all combined through its 'presentational immediacy', an 'immediacy' that appears to us through the way the things we perceive form part of 'an impartial system of spatial extension' (Whitehead 1985: 23). In other words, geography, what Whitehead refers to as the 'spatial relatedness' that underpins the 'presentational immediacy' of things (ibid.: 22), is implicated in the synchronic illusion.

How society responds to the synchronic illusion has not remained constant. The earliest societies denied history by denying change, the present always being treated as if it were the past, present and future all in one. Later, traditional societies tended to admit history but only by compressing it into a single past which they contrasted with the present (Pina-Cabral 1995: 715–35). For reasons outlined in chapter 5, this notion of a compressed past has now given way to an interpretation that is more equivocal. At one level, it can be argued that modern society still sees the world in terms of its 'presentational immediacy', equating its geography, or what is around it, with what is of the present. Yet at another level, there is a deepening awareness of the present's diachronic nature, of how the things that appear to us as spatially coexisting at each moment are, in a sense, located differentially in time, each expressing a different time constant. Paradoxically, the current academic stress on how society reproduces itself through everyday social practice, with all the conditions necessary for its full reproduction being accessible to everyday social practice, is only feasible as a working concept if we treat the present in a wholly synchronic way. Just how far this mishandles the problem of inertia can be shown by a simple analogy. Reproducing the past synchronically is like producing a xerox or photograph of a grand master, one that reduces all its stages of formation to a single instance. To reproduce it diachronically would be to go back over all the successive stages of composition, using the same draught outline, materials, colour mixes and conditions of working in an exact *time–space realization* of the original, reproducing it in the same order on the canvas. Clearly, what is reproduced from the past through each instantiation has some of the limitations of the xerox. Its diachronic qualities are suppressed or glossed.

There is a simple economic reason why society will always be inclined to reproduce the past in a way that glosses its true depth of character. Once viable structures, systems, codes, etc. have formed, they will be carried forward tacitly, not redeveloped. They will still have a learning and maintenance cost, but not a research and development or construction cost. After all, society does not knowingly seek to re-invent the wheel with each generation. New forms or extensions might still be researched and developed, others might emerge as error copies or mis-translations, but existing forms will be taken as given and their transfer between generations reduced to minimum cost. Even this, though, may understate the economic benefits of inertia in relation to the costs of change. It all depends on how one reads the link between past and present, or between each moment in time. If one believes in a visionary thesis of human nature, then one might argue that the Henry Fords of the world are blessed with sufficient foresight to read what will be successful at the outset. The research and development costs involved are merely those attached to Ford himself. However, if we start from the premise that we live in a world of such uncertainty that we only know what will be successful as different strategies and solutions are sifted and sorted by experience, then the calculation of the costs involved changes dramatically. If, at each moment, we do not know what structures, codes, institutional forms, etc. will turn out successful, even just viable, then the costs of change will be the total search and development costs of *all* the strategies and solutions still viable at each moment in time, that is, the social costs of the knowledge involved not just the private costs. Seen in this way, what needs to be carried forward inertially will involve huge sunken costs that are social as much as private. Philosophically, we can argue that all knowledge of the past no less than that of the future exists only through our construction of it via what Shackle called the 'moment in being', a moment whose prime distinguishing feature is its 'solitary' nature (Shackle 1967: 13–15; 1961: 16). Strictly speaking, neither the past nor the future can exist outside of the present. By this is meant that what we read as the past consists solely of evidence and information known to us through its survival into the present. Likewise, what we can say about the future must necessarily be seen through the evidence and information of the present. Such observations, though, can hardly be the last word. To say that society exists only through the 'moment in being' does not mean that it is constituted wholly in the present. It would be an utterly empty moment, devoid of all that can be labelled as human culture, if we suppose that it is constituted solely in the present. As Shackle himself intended, the 'moment in being' only has substance, a consciousness, if we allow it to be enriched by the archaeology of past experience and what some have called its memory traces, either at an individual level or as traces present or emergent within the socio-cultural system at large. Whitehead was in no doubt as to the backward-leaning nature of the present. 'The [wo]man-at-one-moment', he wrote, 'concentrates in [her]himself the colour of [her]his own past, and [s]he is the issue of it' (Whitehead 1985: 27). What is more, the '*how*' of the present 'must conform to the *what* of the past in us' (ibid.: 27). For Sahlins too, society carries what he described as the 'penalty of a past within the framework of which, barring total disorganization, it must work out the future' (Sahlins 1977: 227). As Wilden's quotation at the head of this chapter makes clear, we can extend this point by drawing a distinction between hot and cold societies, with the memory traces recorded by the former physically embracing 'the world outside', its landscapes, whilst the latter 'is more nearly WRITTEN IN ITSELF' as a purely mental trace, though Wilden would be the first to acknowledge that such societies do use the natural landscape as an *aide-mémoire* (Wilden 1972: 408). In both cases, though, 'what is already made becomes a determinant of what is in the making' (Whitehead 1985: 46).

Yet arguably, we need to go further than Wilden's distinction. To use his own terminology, we need to recognize that the different constitutive forms of society – its cultural symbols, organizational forms and technical capacity – all involve different processes operating through different circuits of relations. Each of these circuits involves different logical types that have different processes and different time constants or turnover times (Bateson 1980: 127–43). Further, modern society is increasingly conscious of how such differences transform the outward synchroneity of the present into a far richer diachrone-ity of experience. Such diachroneity, with its acknowledged and powerful role for inertia, can be developed in three critical ways.

First, and fundamentally, large areas of what we would label as culture are comprehensible only as a richly diachronic system of codes, symbols and meanings. The core of what is inherited or passed inter-generationally is com-municated as a given system, as if it were a single, coherent body of norms, values and practices. *Ab origine*, though, it is a heavily compounded aspect of human culture. Typically, it is made up of structures of thought, values, beliefs and traits that have been accumulated over vast lengths of time, including some from the moment when nature first started to give way to culture, all bolted together syncretically.

Once established as viable or useful, a vast body of mundane cultural knowledge is reproduced only in the sense of being retaught and reused, not in the sense of being recreated. In shifting from being knowledge that is created reflexively to being tacit knowledge, accepted implicitly or without question, its status changes. In being carried over as knowledge that does not need to be rethought, it becomes, like Bourdieu's habitus, the basis for 'deeply embedded generative principles' (Bourdieu 1977) that serve to structure every-day life. Further, the conclusion to be drawn from the work of writers like Todd (1985) and Berry (1989: 1–18) is that for all the changes wrought by modernity, there still remain deeply ingrained norms, values and traits to which different 'modernized' cultures are firmly wedded.

Language, for instance, with all its many forms, codes and nuances, is ini-tially acquired or learnt as a mother tongue. Specialized codes and forms may be added later through formal education and training, but the basic form of language is acquired as part of acculturation. In this sense, it is acquired as a purely synchronic system of communication. In the hands of a linguist, though, language has an archaeology, whose different layers of change and adjustment can be recovered and reconstructed. As Saussure especially has argued, this embedded history means that whilst language can be looked at descriptively, that is, in terms of how it operates as a functioning system of signs and oppositions in a given social context at a given moment in time, we can also explore it as a diachronic system by deconstructing its elaborately composite and evolved character (Saussure 1983: 99–181). As with other aspects of culture, what this means is that language is learnt and reproduced as a synchronic or functioning system but, in essence, it comprises a historical system. This is not to deny that it is marginally transformed and extended in range through routine usage, but we can hardly maintain that each 'repro-duction' recreates the conditions which have shaped its character *in toto*. To return to my basic point, the former is a synchronic duplication of a diachronic system, whilst the latter would require a recovery of its history through a genuine diachronic reproduction. Here surely is what Wilden meant when he said that all civilizations are caught between synchrony and diachrony (Wilden 1972: 410).

The diachroneity surrounding society's organizational forms – whether constituted socially, politically or economically – provides a second way in which inertia forms an essential and unavoidable part of all societal systems. Quite apart from the wide range of different time constants that exist between them, with different organizations operating on different time scales, all societal forms have memory traces from prior phases of growth embedded in their character at any one moment. Such diachroneity owes much to their standard growth strategies of differentiation, hierarchization and specialization, and the way in which these work to elaborate rather than replace prior forms. Few are bold enough to tear up more than fragments of the script and to begin again and in the case of those organizations that are, there is absolutely no guarantee that the reforming spirit will be sustained into subsequent generations. As Boulding said in relation to organizational change, 'it is fear of loss rather than hope of gain that limits our behaviour' (Boulding 1981: 103; see also, Isard 1977: 6–7).

Of course, organizational elaboration is usually intended to cope with scalar growth and the need to handle, store and process more information following the expansion of members, interactions and functions, or when faced with extra challenges in their environment. With elaboration, the day-to-day social practices that energize organizations become more specialized. The very terms differentiation and specialization embody processes whereby individuals become distanced from the total information, the so-called social stock of knowledge, which is needed to reproduce institutions and organizations even in terms of their appearances or functioning so that individuals are less able to transform the whole from the standpoint of the individual alone. Not only do these processes of elaboration involve the differentiation, specialization and hierarchization of functions, but also, of codes, roles and rules. The increasingly specialized and selective way in which individuals are interfaced with organizations is one of the factors behind Habermas's notion of the legitimation crisis in modern society (Habermas 1990: 23–44 and 336–67).

A comparison with the less specialized world of hunter-gatherer bands is instructive here. Even at this level, there are specialized roles as regards domestic activities, food procurement, cooking and rituals that differentiate society along the lines of gender and age. Yet far more so than with more advanced societies, it can be said that everyone is involved as participant or observer in the economic, social and ritualistic 'reproduction' of their world. With the development of more complex societies, this immediate involvement in the reproduction of a society's total worldview or cosmology has been greatly diminished. In modern societies, individuals are touched by only a fragment of the total knowledge needed to 'reproduce', let alone re-invent, the world in which they live. We can make the point in a simple contrast. Marooned on a well-stocked island, a hunter-gatherer would survive comfortably. He or she would know of relatively few things but would be sufficiently proficient in most of them to survive. Set down in a typical deciduous woodland or tropical forest, the average hunter-gatherer would have knowledge of many different species or types of animal, nut, fruit, leaf or tuber that are edible. By contrast, an inhabitant of modern society would struggle. He or she would know of many things, and no doubt could imagine the outcome of many sophisticated technologies, but would actually be capable of practising very few, if any, on their own. After all, even in Matthew Boulton's pioneering Birmingham factory, a model of its kind in the eighteenth century, it took nine separate stages of production and nine different operatives to make a simple nail. By the time we reach the early twentieth century, Henry Ford was reportedly able to boast that one car alone had 4,719 parts, and presumably many more operatives to make the different parts. For modern societies, this asymmetry between what organized behaviour can achieve collectively, its emergent possibilities as a social stock of knowledge, and what the individual can achieve if left simply to his or her own devices, has become a powerful source of inertia.

The ability to cope with vastly increased amounts of information has become the hallmark of more recent organizational changes. It has been estimated that 90 per cent of all the human knowledge and codified information created since the last Ice Age has been created in the past thirty years and that the amount will double in the next fifteen years (Haeckel and Nolan 1993–4: 2). The basic point being made here is an important one. Society's rate of knowledge-generation has escalated dramatically in recent decades. But what makes it impossible to put a real figure on its rate of growth is the fact that many of the databases being compiled, serial lists being produced and books being published are diachronous in character, based on historical systems of knowledge that are simply being carried forward as incorporated or tacit foundations. Most knowledge sets are contemporary not in terms of their substantive content, but in terms of their last update or revision. Stephen Jay Gould's broad suggestion that only when a book is 100 years old can we judge its real worth to society makes my point another way (Gould 1996).

At issue is not just the accumulating stock of knowledge, but how it affects the nature of organizations or those social forms that are increasingly designed to gather and process it as a basic 'raw material' (Castells 1989: 13). The knowledge value of goods has become greater than the value of the resources actually invested in them (Myer 1993–4: vii). In fact, nearly 50 per cent of the American economy, and nearly 50 per cent of the value of all its manufactured goods, is now debited to the costs of knowledge-gathering and other information-based services (Haeckel and Nolan 1993–4: 4–5), though,

and not without significance for the argument of this book, only a small proportion (2.7 per cent in 1983) is under the heading of 'research and development' or qualitatively new information (Hall and Preston 1988: 235). Much is
simply 'symbol manipulation on the basis of existing knowledge' (Castells
1989: 15). Yet notwithstanding this qualification, the memory- or informa-
tion-management aspects of the modern organization, enhanced by new
communication technologies, is bringing about a change in its character.
Variously named as the 'learning or intelligent organization' or a 'virtual
knowledge network', lead organizations in the post-modern world are now
more flexibly constituted as a result, more adaptable and responsive in behaviour, and able to operate competitively in an 'any place, anytime, with anyone'
environment (Ives and Jarvenpaa 1993–4: 54; see also Castells 1989: 12–17,
19–21 and 31). It can be argued that such developments have reduced the inertial characteristics of the modern organization, or the post-modern organization as some would now prefer to call it. The hierarchy typical of Fordist
management is being replaced by a flatter and much more open system of
knowledge and information exchange, one less emplaced or concentrated
around particular centres of control (ibid.: 59–60). To a degree, the reduced
inertia of such organizations cannot be questioned. For any organization concerned to maximize responsiveness, inertia would be seen as a potentially
inefficient and unwanted aspect of its functional character. Yet most data-
handling systems are, by nature, memory-management systems, concerned
with how codified information can be stored and sustained. Their gains of
flexibility lie in how it can be retrieved for reference or exchange, not in escap-
ing the fact that vast amounts of such knowledge is already inertial by the time
it is stored, processed and acted on (Wilden 1972: 408). When it comes to decision situations, 'every historical moment' still 'has its own being, its own
history, its own inheritance, and its own knowledge complex' so that decisions
are effectively reduced to 'past recapture when it is gone' (Vickers 1994: 14).

As a source of inertia, the capitalization of landscape through investment
in the built environment introduces a third dimension to diachroneity. Though
it is commonplace for the technical means of society to be seen as an extension of its socio-cultural knowledge, it is a dimension which has found only
an incidental place in analyses that emphasize how social practice sustains and
transforms social structure recursively. In order to be reproduced through
social process, structure becomes merely the prevailing 'roles' and 'routines' of
social process not its more sclerotic or institutionalized forms and certainly
not its concrete expressions in the built environment. Such a perspective is not
unreasonable for those concerned with stressing a socio-theoretic approach to
society *per se*. However, this reasonableness hardly holds for those concerned
with how society operates within its material or spatial world. For the human
geographer, what must matter is not society in its unclothed or naked form,
but how it interacts with its material and spatial world through all its available

forms and means, whether those forms are purely social *or part of its technical and practical extension*. Once we admit the latter into our perspective, then the problem of how process interacts with structure alters. Significant areas of the problem now fit uncomfortably, if at all, into a model that sees process and structure interacting either continuously or recursively. Almost by definition, what is excluded from such a model has inertial properties simply because it is that part of a societal system that is no longer open to easy or all-encompassing revision through everyday social process.

This last point needs expanding. In theory, the built environment can be recreated with each generation or investment cycle. In practice, though, change takes place in a marginal or incremental way. For the economist, decision moments are invariably decision margins (ibid.: 203). More to the point, much of the change that does take place tends to add more to the system than it actually replaces (see, for example, Adams 1970: 37–63). It is physically new stock rather than a substitution for old, though its capacity may replace that of existing stock. As with organizational sources of inertia, the spread of price-fixing markets has brought the diachronic character of the built environment, its variety of vintage, into sharper focus. For this reason, it need hardly surprise us to learn that economists and economic geographers have devoted much more thought to how inertia might be factorized than have social theorists or social geographers. The diachronic character of the built environment, though, is far from being straightforward. In so far as the built environment represents a physical storage of invested energy or capital, then all aspects have a potential for inertia simply because they have a value. However, for all their acute but destructive sense of what is past, it is also the case that free markets have a fairly short memory. Despite the heavy capitalization of the built environment, much of this investment is heavily discounted or depreciated through time. Whatever the residual exchange or use value of the built environment as an inherited form, the present finds a way to free-ride on it. To recap on the point made by Hicks, if we really put a value on all the improvements made to land from the first clearance of the wood or carving out of a routeway, or its potential Ricardian rent, and then allow for interest at a compound rate, the debt of the present to the past would overwhelm all analyses. Not surprisingly, free markets do not have such a long-term memory. However, they are extremely sensitive to the vintage of stock within the built environment over the short or medium term because of the effect which obsolescence may have on output. In a world of unchanging technology and process, phases of investment in new plant and stock would lead to average rates of return on investment, or equilibrium. In reality, though, new phases of investment bring with them new technologies, methods and production functions so that new investment is constantly creating disequilibrium and the greater profits that come from 'best practice'. Seen in this way, mobile capital has a vested interest in creating disequilibrium by maximizing and then maintaining the gap between past or inertial forms and new 'best-practice' forms. It creates obsolescence and, in this sense, inertia, through its investment in what is new, but it does not have an interest in allowing such differential – and, therefore, the essentially diachronic nature of the built environment – to disappear from the system as a whole. In fact, capital will always have a self-interest in the persistence of inertia and the different time constants involved within the system as a whole.

Conceptualizing inertia

There have been few attempts to conceptualize how the inertial properties of socio-cultural or societal systems may affect change. Given the way in which inertia is manifest through a diverse range of values, norms, routines and structures, the prospect for conceptualization might at first sight appear limited. As a prior condition for any such conceptualization, we would need to reduce it to a common form or meaning. One characteristic which all forms have in common is that they constitute information. Seeing what societies carry over in time, from one generation to the next, as information is not a novel suggestion. The archaeologist Clarke put the case for seeing the activities, artefacts and traditions of all socio-cultural systems as being in essence forms of information (Clarke 1968: 101). Likewise, the cultural geographer Wagner suggested that cultural and historical geography can be seen in terms of the information being conveyed across time or between generations (Wagner 1978: 9). Such ideas provide a ready-made basis for a conceptualization of inertia. Like all such ideas, what matters is how we actually develop it as a working concept.

A number of basic issues have to be addressed if it is to have any value for understanding change. The first concerns the kind of information that is communicated across time. In treating it as the defining values, codes, processes and organizational forms of a society, as well as the information that is hard-programmed in the physical structures developed by it, we are shifting the nature of inertia from being something marginal to society, the hallmark of a peripheral niche or group that has failed to keep up with change, into something that has a more central role in how society functions at its very core. In these circumstances, the information sustained inertially comprises the vast sub-stratum of tacit or embedded information by which any society conceives of itself in time simply because it is the 'prior' information which comes to it as given or received. It is this flow of information through time that White captured by his definition of culture as 'the organization of things in motion' (White 1959: 38).

Yet the information which passes between generations always involves a varying degree of selection so that what actually persists or survives is a constrained subset of what actually existed. In the dry language of the information theorist, there is a hypothetical product space that always embraces a far greater range of possibilities than the more limited range of what actually existed (Ashby 1968a: 109; Rapoport 1968: 141). Inertia can be regarded as shaped by the constraints on what is communicated. Moreover, the very existence of such constraints acts historically or through time to create organization or negentropy (Ashby 1968a: 109). This is why we need to see societal systems as essentially diachronous, as having inertial properties, rather than as simply synchronous. Geographers are inclined to see order in any form as something not just spatially deployed but as spatially constructed, irrespective of time, yet, in essence, it is something that necessarily *unfolds* through time as much as space (cf. Bourdieu 1992: 136). What ensures the existence of order is the continual and sequential exercise of constraints on what is viable so that, through time, order emerges as something negative, the expression of society's conservative bent. The extent to which geographers have recently stressed the way in which the history of a place is written into its character, and works to constrain what can subsequently happen, makes this very point.

To assert that inertia is about the communication of such information across time still leaves us with the question of what this communication actually involves. Like all exchanges of information, it is not a one-to-one transfer. As Bateson put it, though the message is a mapping of what is being conveyed, the map is never the territory (Bateson 1980: 37–8). One is a transform of the other. Moreover, as a communication of information, there need not be a transfer of energy in any direct sense (Bateson 1972: 416). If we take the codes and rules stored in a system of language, then their acquisition involves an expenditure of energy by both the person(s) teaching and the person being taught, but there is no meaningful sense in which we can say that the learning of language by the latter involves a transfer of the energy expended by a society in developing a language, though patently it does involve a transfer of information. We could apply this to many other key areas of a society and its landscape, from organizational forms to ways of life.

Of course, not all information accessible to a generation is transferred to subsequent generations. A choice is involved. Two fundamentally different types of selection exist. The first is the initial creation of information through the development, intentionally or unintentionally, of new or variant codes, strategies, forms, structures and solutions: this is a creative, divergent process, one that leads to diversity. It opens society up to what some would call 'noise, the only possible source of *new* patterns' (ibid.: 416). It is a process well captured in the work of Allen on 'evolutionary landscapes'. Change in such landscapes, especially the qualitative change that lies at the heart of evolutionary models, requires the noise or new information produced by 'the error maker who can move uphill', the person who breaks ranks (Allen 1988b: 107; see also, Bateson 1980: 57). These error makers are stochasts who do not respond 'to the information about the present returns on effort' (Allen 1988b: 117) even though 'at each and every instant it would be better not to make errors, since the majority of these are loss-making' (ibid.: 107). Where society is able to absorb such deviation or noise as part of its ongoing selection, it leads to morphogenesis and real change (Wilden 1972: 331). The second is the selection made by the sifting and sorting of what is viable or successful from what is not: this represents a convergent process, one that works to conserve dominant forms. Because this second process adds nothing that is new, we can see it as working towards a principle of morphostasis (ibid.: 331), maximizing the potential redundancy or predictability of the information that is carried forward by society across time (Bateson 1972: 412). Inertia is bound up with this second process of selection and involves a reduction of the possibilities opened up by the first process.

A further point worth making here draws on my earlier argument regarding the so-called synchronic illusion. In origin, rules, codes, forms, processes and organizations, together with the built environment, are intrinsically diachronic, developed over varying lengths of time. Their communication, though, patently simplifies the problem by treating them as synchronic, that is, as a given functioning system so that only their appearances or relations are effectively communicated. By relying on this inheritance, by being conservative and 'duplicating' rather than creating many of its rules, codes, forms, processes, etc. afresh, society saves on the costs (search, replacement, etc.) and risks (failures, cul-de-sacs, etc.) involved. There is a strong case for arguing that if a society did have to recreate its different forms of being and expression *de novo* with each new generation, then the development costs would overwhelm it. In this sense, society might be said to free-load on its past, accepting much of what is bequeathed to it as a cheaper and less risk-laden deal. Stated as such, inertia becomes a calculated compromise between what persists and what is allowed to change.

Arguably, the choice between them is divided by a trough created by the declining utility and benefits of old choices and the initially high costs and risks of new ones. Construed as a single system of choice, we might derive models which, as Isard once proposed, try to calculate the trade-off between maintenance, redundancy and novelty in exact terms (Isard 1977: 9). In reality, though, societies are simply not organized in this sort of coherent or homogeneous way. As already implied, a selection for maintenance or morphostasis draws on a process of convergence, whilst that for novelty draws on a process of divergence (Bateson 1980: 191). Arguably, these differences are best rationalized by seeing them as constituted socially, expressions of different groups, sub-groups or vested interests operating in different spaces. Within any society, some will have cause to feel that they gain from inherited solutions, maintaining the past, whilst others may have cause to feel that they lose by them.

There is a further dimension that needs to be explored. Society's capacity to derive new rules, codes, forms, processes and structures involves an expenditure of choice. The time and energy given to their formation creates a capital stock in the sense used by Bourdieu but the energy committed to their formation is irrecoverable in the process. Yes, society can set out to devise variants of its rules, codes, forms, processes and structures and it can use old ones in new ways, but it cannot retrieve or recycle the energy already sunk in the development of existing or prior forms. It can only abandon them, or commit more energy to their maintenance as systems of information. This leads to a paradox. The initial creation of rules, codes, forms, processes and structures can be seen as a positive act, one that creates the possibility of new societies and new landscapes. In the process, though, it also creates cultural landscapes of expended choice, of *constrained* variety, because it commits what were free-floating resources, both human and physical, to fixed or bounded forms. Cultures, declared Said, 'impose corrections upon raw reality, changing it from free-floating objects into units of knowledge' (Said 1991: 67). As Rowntree and Conkey made clear in a discussion about the symbolizing of landscape, to turn what is 'free floating into units of knowledge' in this way is not just to 'store information', but equally to 'constrain' or 'restrict its flow' (Rowntree and Conkey 1980: 461). Yes, if a society still has what Bateson called 'unused freedom', it can go on expending choice by creating new forms of knowledge or adding to existing ones, but once 'free-floating objects' have been transformed into 'units of knowledge', then they cannot recover what has already been expended. For this reason, the pioneer creation of landscapes may lead to new forms, but it does so at the expense of flexibility. Seeing human geography in this rather negative way alters our perspective. It introduces the possibility that change can now be framed as a study in what Bateson called an economics or budget of flexibility (Bateson 1972: 508) and how well societies are budgeted as regards their free-floating resources, whether cultural, spatial or material.

Landscapes of constrained variety and the geography of change

A common misunderstanding is to assume that the existence of capacity in a society gives it an enhanced ability to cope with change. In fact, larger, more complex societies face the greater difficulties: the 'more adapted a culture', maintained Sahlins, 'the less therefore it was adaptable' (Sahlins 1977: 221). Arguably, this is because such societies invariably have a greater burden of prior information to be carried forward, more that is specialized and more that is bound. Many of the major innovations underpinning the development of more complex societies address the problem of how information can be communicated, stored and retrieved: language, mnemonic-based memory systems, the use of ideograms, alphabetic writing, the printed word, indexing, hierarchical or tree-based classificatory systems, and electronic data handling with its profound shift from the recording, storing and processing of analogue data to the potentially more efficient form of digital data. All such innovations have released extra potential for information storage and communication. However, they represent only part of the way in which information in the wider sense is stored and communicated. Increasingly, as society has stored information outside of itself, it has created landscapes rich in expended choice. The more such landscapes have been symbolically coded, the more they have been used to differentiate and structure society itself in an instrumental way, and the more they have been used to articulate and sustain society's increasingly extended technical and physical capacity, then the more such choice has reinforced consecutiveness and persistence.

Ordering information geographically is unquestionably deep-rooted in society. Many early societies used simple maps not only as a representation of geographical information but as an *aide mémoire* for cultural information generally (Lyle 1993: esp. 64–9). Perhaps the most celebrated example is the aboriginal use of dreamtracks (Head 1993: 489–90; Eliade 1973). In classical Greek society, professional remembrancers also ordered information geographically as a means of remembering it, linking particular types of data to particular spaces as the basis for its recall in a structured way. This particular tradition is thought by some to have led on to a 'western preoccupation with visual and spatial root metaphors of knowledge' (Fabian 1983: 109). Even today, it is commonplace for 'western' knowledge and the social boundaries that demarcate it to be seen geographically, as a map of knowledge (Ziman 1985: 7 and 10–11). In fact, there is a case for supposing that all societies have made use of 'spatial root metaphors' to order the way in which they not only envision the world but knowledge generally. As an extension of such metaphors, all societies have also tended to use their landscapes referentially, or to signify basic beliefs and ideologies (Duncan 1990). This is reinforced by the growing way space has been used in modernizing societies to order a wide range of basic information geographically, from property rights to census data, from legal and political jurisdictions to cultural identity. Space, or the environment around us, is – as Wagner has pointed out – 'suffused with [wo]manmade symbolism, peerless and imperishable repertoires of the past experience of all the species', a symbolism that acts as an extended information store (Wagner 1972: xi).

We can apply the same thinking to the organizational forms of society. Their increasing spatial differentiation and the differentiation of higher and higher levels of control, including the differentiation between line and staff, has enabled organizations to cope with more information. In so far as significant amounts of this information can be regarded as specific to the organization and its particular history, it effectively amounts to an investment of energy in maintaining a flow of information through time. Olson's suggestion that given time and stable boundaries of operation, organizations become increasingly collusive, organizing themselves into distributional coalitions that are more concerned with how wealth is distributed rather than created, can also be transposed into an argument about how they invest still more of their available energy in seeking to persist as given forms rather than in creating new information, the former having lower costs and risks than the latter. Significantly, their persistence is geographically mapped (Olson 1982: 40–1, 70–1 and 98).

Likewise, the fact that complex and advanced societies have a great deal of past information stored and communicated in the form of the built environment is beyond dispute. More to the point, it represents an expenditure of choice which, as bound information, has a high level of fixity. Except in special circumstances, such as following war damage, fixed capital forms have shown a resilience to easy change. As in Watson's study of Halifax (Nova Scotia), the geography of many western cities is as much about 'relict' spaces of accommodation, retardation, ossification, deviation and deterioration as it is about spaces of change (Watson 1959: 110–43). Given contemporary society's heightened sense of what is past, of diachroneity, any element of the built environment that manages to persist long enough, even if it has largely persisted through neglect, now acquires the status of heritage or symbol, and a near absolute resistance to change. Even in Britain, where we can cite many exceptions, such as when thousands of slum houses were cleared over the mid-nineteenth century to make way for the new London railway termini, the comprehensive redevelopment schemes which redrew parts of Britain's major cities during the 1950s and 60s, the ongoing change in high street frontages, or the radical redevelopment of London's Docklands over the 1980s and early 90s, the overall tendency has been for the forms of the built environment to persist long after the disappearance of the social and economic conditions that first brought them into being. Further, the growing recovery of the past as a diachronic experience has meant that pressures for conservation and gentrification have grown not lessened. Perhaps more so than at any other time, the built environment is expected to order information as much through time as through space.

This growth in the amount of information being carried forward by society patently amounts to an increase in the inertia of landscape. As a force in the making of geography, inertia acts through the conditionality which it introduces into how information is carried forward. As information, it may be reduced and transformed in the process, but in its effects, it serves to constrain outcomes. To put this another way, any durably installed order possessed by landscape derives from the way inertial forms serve to foster repetition, a consecutiveness of form, thereby reducing the information needed to that which helps to maintain the essence of what is being communicated in its simplest form. Though this order appears at each moment to have synchroneity or simultaneity, this does not alter the basic fact that it is an order whose essential forms and relations are developed through time as forms or relations of succession. Yet whilst it is an order which, first and foremost, is constituted temporally, it is also configured geographically. Indeed, if, because of its essential consecutiveness, we see inertia as a largely irreversible expenditure of choice over how free-floating resources are used, then it also follows that inertial landscapes are landscapes of constrained variety, landscapes across which choice over how such resources are used has been expended in a geographically differential way. The point which I am working towards here is that in a world ordered around the economics of flexibility, landscapes become overwritten by a geography of flexibility/inflexibility (Bateson 1972: 507–10). Where societies, or groups within them, have expended a choice over how the flexibility available to them should be used, a choice which is invariably emplaced in the landscape, then it produces a geography of inertia that constrains subsequent choice.

Once we see it in these terms then the problem of change alters. Contrary to those who have rejected the idea that the way forward lies in understanding how change and stability are related, there is a case to argue that they are related provided we grasp that their direct confrontation can be displaced through a geography of flexibility and inflexibility. Logically, change has a greater likelihood in those areas where there are greater resources of 'unused freedom', that is, greater resources of flexibility or plasticity about how available choice over the use of resources might be expended. Whereas areas held firm by the stasis of inertia have an experience or history of change that is necessarily convergent, reinforcing consecutiveness, areas which have greater reserves of 'unused freedom' are, by contrast, those which combine a greater capacity for divergence, novelty and macroscopic diversity, with the flexibility that could feed runaway growth. Seen in this way, the geography of flexibility and inflexibility provides us with the foundation for a geographically informed concept of change, a structural reason for supposing that change may be more likely in some areas than in others through the way inertia maps constraint and, as a consequence, creates a geography of relative opportunity. To understand this geographically informed concept of change, we need to establish how society or groups within it exploit space strategically so as to circumvent inertia, exploiting instead those areas that have greater residual flexibility over how choice might be expended.

9

The geography of societal change: a concluding synthesis

There is no such thing as social change in general. Many large-scale processes exist; urbanization, industrialization, proletarianization, population growth, capitalization, bureaucratization all occur in definable, coherent ways. Social change does not.
(Tilly 1984: 33)

... hybridity to me is the 'third space' which enables other spaces to emerge. This third space displaces the histories that constitute it, and sets up new structures of authority, new political initiatives, which are inadequately understood through received wisdom.
(Bhabha 1990: 211)

In this final and concluding chapter, I want to resolve the question which I posed in chapter 2 of why societal change needs to be approached through a clearer understanding of the inertial properties of socio-cultural systems and the intrinsically geographical dimension which this introduces to the problem of change. Two aspects of the problem deserve to be restated. First, the more complex a society, and the more stable, then the more likely it is that a greater amount of its available energy is expended on maintaining the flow of given or received forms of knowledge, that is, on inertia, rather than on the development of new forms and possibilities. As an exercise in the care and maintenance of bound information or expended choice, this consecutiveness of choice serves as a constraint on subsequent change. Furthermore, it is a constraint which has a strong geographical component to it so that the residual opportunities for change, the balance between the resources of flexibility and inflexibility, can be seen as a geographically framed problem.

The second point which needs to be restated is that, quite apart from variations in the geography of flexibility and inflexibility, fundamental differences of attitude exist, both between and within societies, over whether particular forms of socio-cultural knowledge, values or structures should be maintained or abandoned. These differences permeate all dimensions of social life, from its most deep-rooted norms to the more manifest structures of the built environment. In most analyses, such differences are couched in terms of the oppositional differences between the values and self-interest of those who make up the core of a society and of those who make up its periphery. Such binary or polar differences can be expressed relatively as well as absolutely, so that there are those within the marginal spaces of the core who see themselves as marginal just as there are those within the periphery who feel bonded to the values and structures of the core. Though these differences are developed by a range of writers (e.g. Johnson 1966: 67; Shields 1991: 3), I want to re-emphasize the point through the thinking of Shils. Every society has a centre, he argued, formed by the 'centre of the order of symbols, of values and beliefs, which govern' it (Shils 1975: 3–4). However, and this is the essence of his argument, it is not the whole of the values or beliefs within a society (ibid.: 3). Other, potentially rival, values and beliefs exist within its frame. Patently, Shils is simply saying that all societies have groups within them who gain from upholding its wider values and beliefs whilst others have a vested interest in upholding alternative norms or in questioning the prevalence or pattern of power and rewards within it. Though his language might be seen as mapping such differences, referring variously to the centre as a 'zone in the structure of society', an 'ecological domain', he rejects the idea that it is a 'spatially located phenomenon' or is connected with geography (ibid.: 3). It exists only in a phenomenological sense.

I suspect that in the light of what is now being written about the spatiality of social life, he might revise this verdict. Yet even amongst those who construe the spatiality of social process in more explicit terms, the growing stress on imagined geographies that are states of mind without also being tangible on-the-ground expressions of geographical structure, might suggest that Shils's wholly phenomenological mapping of the problem is all there is to say. We are led to believe that societies and their constituent groups think geographically, spacing their differences mentally, but they do not necessarily practise it. We can see this best in the emergent debate over third space. As defined by Bhabha, third space is composed of the hybridity that comes from being betwixt and between, and the differences which it generates as regards signification. Those living in third space are constantly negotiating across these differences in signification, translating their meanings into new hybrid forms (Bhabha 1990: 211). In effect, third space is 'liminal space, in-between the designations of identity' (Bhabha 1994: 4). Given Bhabha's stress on how the spread of symbols and the time displacements involved is a source of these differences in signification, we can hardly doubt that Bhabha sees the hybridity of third space as underpinned by a form of geography or space that is intrinsic to it.

To a degree, Soja's biographical reading of Lefebvre's ideas leads us to a similar conclusion. Lefebvre's family roots in the Pyrenean margins of the French state are juxtaposed to his academic life in Paris. Soja reads Lefebvre's 'hybridity' as deriving not just from the 'third space' formed by his marginal roots, between what was within and without, but also, as deriving from his constant mediation between life in the provinces and life in Paris, his 'ambidexterity' in negotiating their differences (Soja 1996: 30). Although such a notion of third space is different from that defined by Bhabha, one being about the interstitial hybridity constituted within marginal spaces and the other about the external hybridity, the hybridity beyond, forged between them and a core area, I have no real disagreement over Soja's reading here. Each generates a form of 'liminal space'. However, Soja uses Lefebvre to develop a somewhat detached view of third space, one with metaphorical rather than metonymic meaning. Indeed, in a later discussion, he claims that the essence of globalization and post-modernity is their 'disordering' effect on space and the way in which the great conformities of space are being broken down into a multiplicity of forms and meanings, at least in the city. Third spaces, with all their new orientations and imaginative possibilities, are seen by him as an outgrowth of this disordering (ibid.: 61).

There is a danger here that we reduce geography solely to metaphor, constituting it as a detached, imagined form. The whole point of this book has been to show that geography matters precisely because it embodies the materiality of social life. Society variously symbolizes space, organizes it and enhances its value through massive investments in the built environment. To put this another way, imagined geographies are not thought out in the abstract but are simply different ways of codifying or symbolizing what is ultimately real. However much we dislocate or 'disorder' space, its 'third spaces' still need to be reified as real world spaces if they are to be open or flexible spaces, whether in the form of the Pyrenees in the case of Lefebvre or 'the other side of the tracks' in the case of hooks. To return to Shils, there can be no meaning to structural positions like 'centre' and 'periphery', just as there can be no meaning to notions of hybridity as a self-standing concept or wrapped up as third space, if they are not also seen as spatially meaningful in a practical, everyday sense.

If we accept this and if we align such differences with the geography of inflexibility and flexibility expounded in chapter 8, there are good reasons for supposing that core groups will occupy the spaces that are potentially more inflexible simply because they consume and invest more of a society's real wealth, focus more of its organizational needs around themselves and consume more of its symbolic capital on reinforcing their position. In other words, not only can it be said that those who gain most from a system are less likely to will change to it, but they are also more likely to be surrounded by constructed spaces across which more choice has been expended and on which more is being expended to maintain inertia, or the consecutiveness of choice. Likewise, we can expect marginal or interstitial areas, spaces which are not sites of domination, to be occupied by those who gain least from a given system and whose spaces of identity have, relatively speaking, greater resources of flexibility or unused freedom still available.

Once the pattern of centre and periphery, will and opportunities, domination and resistance, is mapped in this way, then it opens up a geography of change. Such a geography discards assumptions about how advanced or complex societies are best organized for change because they have adjusters as well as regulators, that is, sub-systems which monitor for opportunistic change as well as for stability. As a number of studies have demonstrated, the more organized a society then the more problems it will face over change, having both greater amounts of inertia to sustain and greater vested interest in defending the *status quo* against radical change. Their capacity for movement is eventually inhibited by their own weight, reaching a point at which the sheer cost of expanding or even maintaining the *status quo* overwhelms the freedom needed for adjustment. Once we pose the problem in these terms, then it follows that change will not work itself out in a uniform way. The differential distribution of those who might gain most from change, together with the differential distribution of inertial forms and unused freedom, combine to produce a distinct geography of change, one in which the accumulated past, or the history of the system, matters.

Enlarging on this geography of change, I want to organize what I say around three themes. First, I want to consider how the balance between three standard forms of societal change – involutionary, revolutionary and evolutionary change – is affected by the way in which the geography of inertia and unused freedom is contoured or mapped. If inertia and flexibility are geographically emplaced aspects of society and its spatiality, it also follows that a society, or groups within it, can develop a strategy of change by exploiting the greater flexibility present in some areas compared to others. As a second theme, I want to explore this point further by looking at ways in which societal or socio-cultural change has exploited space strategically. My third and concluding theme will round off my argument by drawing out the significance of inertia for how society can handle change in the future.

Change and the geography of unused freedom

Involutionary change

Involutionary change, the shift towards a more elaborate internal form, affects all societal systems. In a sense, it represents a solution that maximizes the inertia of given forms. It emerges as a singular form of change when pressure for continuing change exists in a core area but where there is little opportunity for directing such pressures into new areas of unused freedom. Such areas can be heavily constrained in their response by the social costs of carrying over inertial forms of socio-cultural knowledge and organization and by the fact that dominant or elite groups have a vested interest in preserving the system as it stands. In these circumstances, pressure for continuing change *in situ* is strongly conditioned by what already exists, or by Unger's pull of sequential effects. Change becomes more of the same, with scalar growth being accommodated through standard organizational adjustments like greater internal differentiation, specialization and hierarchization, processes which act on an already closed agenda. Up to a point, such adjustments can be successful, with each new investment of fixed, organizational and symbolic capital leading to sufficient returns. Ultimately though, as Service and Sahlins's Law of Evolutionary Potential reasoned, with each step towards a more specialized and adapted form, 'the smaller its potential for passing onto the next stage' (Sahlins and Service 1960: 97). Putting the point in more economic terms, Tainter argues that a point is eventually reached at which the marginal return on complexity alone, as the only solution, declines. As a strategy, it ultimately 'yields comparatively lower benefits at higher and higher costs' (Tainter 1988: 127), as more and more resource is devoted to standing still or system maintenance (ibid.: 115) and as more growth-retarding organizations emerge (Olson 1982: 98). Without making radical qualitative change, the mere elaboration of a system towards greater complexity, its involutionary development, has finite limits and can lead to stagnation if not collapse as costs rise against returns (see also, Eisenstadt 1964: 235; Arrow 1974: 49; Boulding 1981: 107). The problem with complexity as a solution is that a great deal of its expended resource is necessarily embedded in a fixed geographical form, as codes, roles and organizational structures and as fixed capital become focused around politically dominant centres. Once densely mapped with patterns of expended choice, the need to sustain the given form of these centres, their ongoingness as the symbolic focus for a co-ordinated system, becomes confused with the very existence of the system itself.

Revolutionary change

Despite its celebration as such, revolutionary change cannot be seen as an open option. By this, I mean that there is no case for arguing that it is a normal path of change, one to be found or used in a variety of circumstances. If we take account of all the different types of change outlined in chapter 1, together with the instances of actual change detailed in chapters 2–3, then, whatever might be claimed by those who think all change is about triggering or preparing for revolutionary moments that can transform a society in its entirety, it appears as only one of a range of different types of change. Moreover, there is a case for arguing that it is one of the most condition-bound forms of change (Touraine 1990: 121–41). In trying to specify what these conditions are, I want to begin by arguing where it is unlikely. If resources of unused freedom, however defined, are still available to a society or its sub-groups, then it is unlikely to seek revolutionary strategies of change. In the first place, all social groups prefer risk aversion strategies so that if scope for the modification or deepening of existing solutions still exists, they are more likely to pursue such a solution. Also, compared with the continuing use and refinement of existing forms, revolutionary change will initially carry heavy redundancy costs for existing forms as well as high risks in the search for better alternatives.

If we turn this into a more positive evaluation of the conditions under which revolutionary change is likely to be adopted, it follows that it is more likely to occur in circumstances in which the degree of unused freedom available for change is minimal and where the possibilities for involutionary change have been exhausted or are subject to rapidly diminishing returns, so that risk-free and low-cost solutions are no longer available. For some, reading revolutionary change in this way removes too much of its chance, contingent nature, but in reply, it could be argued that all events have a wider background context that biases or loads their probability of occurrence and their probability of outcome even though the events that actually trigger them may be entirely random and depend for their precise timing on the actions and motives of charismatic individuals. Further, it needs to be kept in mind that if revolutionary change is only one of a range of different types of change, then its choice or occurrence has to be argued alongside other pathways. It cannot be taken for granted.

Arguably, a logical reading of the conditions under which it is most likely to occur is one which stresses the extent to which a differential geography of flexibility and inflexibility constrains how pressures for change work themselves out, forcing conflict between groups over the use of resources – social, physical and spatial – and over how the rewards of their use are allocated. In a particularly systematic analysis of what constitute revolutionary conditions, Johnson drew the distinction between functionally diffuse societies and functionally specialized ones. The former were typified by a homogeneous spread of values and low interdependence between their regional parts: their low measure of internal difference, with each region being alike in its character, fostered rebellions but rarely revolutions. By contrast, functionally specialized societies, especially those based on regionally differentiated environments and regionally differentiated economies, generated differentiation amongst their subsets and, because of their specialization, high levels of interdependence (Johnson 1966: 146–7). In the case of the latter, Johnson sees the combination of specialization and interdependence as generating a particular tension, with the former encouraging fundamental differences of value and the latter, differences over how the system *in toto* should be managed. Rebellions become a possibility when the different subsets and their communities become 'dis-synchronized', meaning when the values underpinning the system at large no longer accord with the geographical division of labour between the various subsets (ibid.: 147). Put another way, we can see this as a case of different spheres moving at different rates, each paced by its own mix of inertias and flexibilities, but with the overall range of possible solutions limited by a degree of geographical closure about what is possible. The geographical emplacement of these 'dissynchroneities' and their economic geography was for Johnson a crucial part of the argument. Whereas some societies respond to such situations by pursuing strategies in which this sort of potential conflict is deflected or transferred towards external spaces and resources, what is significant about revolutionary change is that it is a form of change that challenges inertia itself, breaking the mould of custom by destroying it.

Any attempt to define the pre-conditions under which revolutionary change might be sparked off in core areas, though, must be tempered by the conclusions reached by Skocpol. In what is still the most comprehensive treatment of the conditions under which revolutions occur, she argued that historically significant revolutions like those of France (1789), Russia (1917) and China (1949) began not through conflicts within their core but as a challenge by *marginal* elites against the values of the core (Skocpol 1979). Significantly, she sees such elites as using marginal space as 'tactical space', space within which there are greater opportunities for thinking differently (ibid.: 195). Latterly, Meinig has added to her list by suggesting that *marginal* elites who served as cadres of the British Empire, Janus-like individuals who could see both ways, led the American Revolution. Of course, in this case, these marginal elites led a revolution that severed their ties with Britain, replacing it with a core of their own making. Like Skocpol, Meinig saw a general lesson to be learnt from this initial eruption of revolution within the marginal areas of empire (Meinig 1986: 373–4). Clearly, whatever case might be made for seeing revolutionary change as more likely to originate within closed core areas, the reality is that some of the most significant historical revolutions were revolutions of the periphery before they were revolutions of the core.

Evolutionary change

In defining evolutionary change, it is important that we distinguish it from developmental change. The changes embodied in the latter represent the playing out of a latent but fixed potential. The conditions which determine this succession are internal to the system and remain unchanged. No extra information is required to produce the different stages of development beyond that present at the outset. By contrast, evolutionary change is the emergence of something that is qualitatively different through the input of something new or novel. However, the idea that societal systems have undergone a process of progressive evolution, or ontogeny, transforming themselves from within through a linear succession of qualitatively different and more advanced systems, finds little support in the debate over societal evolution, despite attempts by some social theorists to define how such an ongoing transformation might be achieved. Indeed, when it comes to evolutionary change, it is doubtful whether there is any value in treating any individual society as such a closely co-ordinated system that it could somehow change *en masse* or as one and could go on changing as such. More realistic is to see societal systems as made up of different sub-groups, sections and communities who are variously rather than uniformly integrated into society, some tightly and others loosely. Those at the core of society are more likely to hold to a uniform system of values, norms and beliefs, and to maintain the ideological symbols and organizational forms that define its core, devoting a great deal of their residual free-floating resources on regulators that ensure inertia. By contrast, the peripheral parts of a society, as well as those societies beyond its formal bounds but who are under its hegemony, are less likely to uphold its identifying values, norms and beliefs or to play a part in maintaining its ideological symbols and organizational forms except through those local elites who may have been co-opted into servicing the needs of the core and through the wealth that might be extracted from such societies and concentrated in core areas.

Applied to the problem of change, we can expect core areas to have both less diversity and a greater interest in maintaining the *status quo*. By comparison, and for a variety of reasons, we can expect peripheral groups and societies to manifest more diversity and hybridity but to have less interest in maintaining the normative structures embodied in the system at large. Of course, peripheral societies may have a vested interest in maintaining their own rival values, symbols and organizational forms, using them as a basis for identity. What matters is how much extra flexibility is gained from the fact that peripheral and marginal areas are usually less burdened by the expended choice which core groups have invested in the symbolic, organizational and fixed capital forms that define core areas, and from the fact that they possess the kinds of interstitial, betwixt-and-between spaces within which one is more likely to find the new forms, variants and combinations that come from hybrid or equivocal identities, error-making and hill-climbing. Put simply, when seen collectively, they have less bound information and more – to use Wilden's phrase – 'semiotic freedom' (Wilden 1972: 233). To recall a point made earlier, core areas are prone to being convergent whilst peripheral and marginal areas are prone to being divergent, that is, to engendering spaces rich in macroscopic diversity. In a world in which conditions and demands are constantly subject to adjustment, to have a pool of diversity from which potential 'error-makers' and 'hill-climbers' might emerge is a virtue. To have monolithic, uniform systems of order and routine, the kind that invest heavily in regulators, or auditors, is not.

These somewhat abstract points can be turned into a more substantive argument by reference to the growing number of case studies that stress the role not just of peripheral individuals, groups and societies in the generation of difference, but also, of the peripheral spaces which they occupy. Whyte's *Organization Man*, a pioneering study, described the way in which large corporations expected conformity of their employees, even amongst those involved in research and development. Quoting Lancelot Whyte, W. H. Whyte's counter thesis was that it 'is usually the relatively isolated outsider who produces the greatest novelty' (Whyte 1960: 201). His point has been heeded. Appreciating that real change amounts to something more than duplication, the director of Xerox's Palo Alto Research Centre has argued that organizations learn on their periphery 'where intellectual outliers, renegades, closet philosophers and others who think "outside the dots" experiment and speculate' (cited in Haeckel and Nolan 1993–4: 11). What is striking about such ideas is the extent to which they find echoes in other fields. In cultural studies, Bhabha's work on the hybridity of marginal 'transnational' spaces is clearly relevant, for these 'in between spaces' provided 'the terrain for elaborating strategies of selfhood – singular or communal – that initiate new signs of identity and innovative sites of collaboration, and contestation' (Bhabha 1990: 207–21). Hall too, sees the geography that may underpin novelty, proposing that 'marginality has become a powerful space . . . increasingly, anybody who cares for what is creatively emergent in the modern arts will find that it has something to do with the languages of the margin' (Hall 1991: 34). The creative capabilities of peripheral spaces also feature in discussions about long-term shifts in political geography with, as Gilpin put it, a 'tendency for technology and inventiveness to diffuse from dominant powers to peripheral states (which in turn become the dominant powers of an enlarged international system)' (Gilpin 1981: 182–5; Mann 1986: 179). Such a conclusion sits comfortably alongside Sahlins's suggestion that 'throughout history, advanced cultures have displayed a special gift for generating further advance, but not so much from their developed centers as on their ethnic borders' (Sahlins 1977: 226). To stress the manifold ways in which peripheral societies and spaces can innovate, though, and to leave the argument there provides a misleading cause–effect model of how evolutionary or qualitative societal change may unfold. In practice, and equally important, is what we can call the ecology of innovation. We need to grasp how particular solutions or strategies, however successful in themselves, may be constrained by their setting, a setting that includes patterns of expended choice and commitments to inertial forms. Again, in so far as peripheral spaces are likely to have greater resources of unused freedom, they are more likely to experience runaway growth when new solutions prove not just viable but successful. Seen in this way, change is not simply about new information, pure and simple, but about how this new information fits with what already is and the extent to which the inertia of what already is – society's gross need to carry forward information as unchanged – affords sufficient flexibility for treating 'errors' or 'noise' not as a short-term deviation but as new information about how things should be done routinely or tacitly (Schrecker 1948: 216–37; Boulding 1981: 107; Nelson and Winter 1982; Porter 1990).

The geography behind societal change

We cannot see these different types of change as operating in an exclusive way. Each has a context in which it is relevant. What matters is to determine the geographical circumstances under which one might have precedence over the other as a strategy of change. If space has a 'strategic function' for society (Gregory 1994: 275), then it can be no surprise that it is implicated in the strategies by which society copes with both inertia and change. Faced with the mapping of expended choice and its implicit geography of inflexibility/flexibility, groups within core areas are just as likely to exploit the greater freedom of choice present in peripheral areas as are groups directly identified with such areas. The former, a core-driven change, can arise from attempts to avoid the constraints present in core areas and societies. More to the point, by exporting or displacing their pressures for change into peripheral areas, they avoid these constraints *without challenging them*. Indeed, the most imperial societies are often the most inertial, exporting their need for ongoing change or for more free-floating resources to the margins of empire. In such cases, their core areas remain inertial whilst peripheral areas carry the costs of change. The extent to which core-driven change seeks areas of 'unused freedom' is also shown by the way groups in a population of societies can push change upwards into the 'peripheral areas' represented by higher forms of political and economic organization. We can treat this upward, integrationist expansion as also being an exploitation of 'peripheral areas' because, in practice, it exploits the unused organizational space that lies between societies. By comparison, change originating wholly within peripheral or interstitial areas tends to develop through their greater indigenous diversity, and their less constrained capacity for producing innovations or combinations that are capable of runaway growth. Further, through this runaway growth, some examples have shown a capacity for exporting this change back into core areas via a revised mapping of what is core and what is periphery.

   We can see how these different forms of change exploit space strategically by looking at them more closely. I propose to do so under three headings: core-driven outward change, core-driven upward change and edge-driven change or change from within the periphery.

Core-driven change: the outward search for flexibility

Logically, core-driven processes are rooted in a problem generated by all core systems, the need to sustain growth. As they mature, they invariably use up flexibility, becoming more concerned with maintaining inertia. As more energy is devoted to maintaining a consecutiveness of choice, then returns on investment diminish and organizations respond by grouping themselves into distributional coalitions. Seen negatively, this can lead them into involutionary forms or into the pre-conditions for revolutionary change as different groups compete over a static or declining unit of resource. Alternatively, some societies can seek to solve their problems by looking to expand their resource base into the areas of 'unused freedom' that may lie around their edges or even beyond their bounds via a spatial fix (Harvey 1981). It is a strategy of change first hinted at by Hegel in his *Philosophy of Right* (1821). Hegel represents 'mature civil society' as invariably afflicted by internal contradictions. What Harvey is concerned to highlight is the way Hegel sees imperialism and colonialism as solutions to these 'internal contradictions' (Harvey 1981: 2). In effect, 'civil society is forced to seek an outer transformation through geographical expansion because its inner contradictions admit of no internal revolution' (ibid.: 2). Harvey traces a further variant of this 'spatial fix' in von Thünen's thinking. Where labour and capital are free to exploit the 'open' conditions of a frontier, then it fosters stability and harmony in the core of a society by allaying its tensions. However, where the frontier is closed, so that its 'freedom' is no longer open to a society, then 'burgeoning social unrest must be countered by an inner transformation of civil society' (ibid.: 5; see also, Boulding 1981: 107; Gilpin 1981: 181; Harvey 1996: 110).

Harvey moves the debate forward by rooting the problem more securely in the general nature of capital. He stresses how the inner and outer transformations of society are linked conceptually through the crises brought about by the over-accumulation of capital within the former and the export of capital to the latter in order to resolve such crises (Harvey 1981: 7–8). That capital moves around within economic systems, exploiting and then discarding the social, physical and spatial resources of different regions when profit rates average out, is now an established feature of the geographical debate over industrial capital. Indeed, Harvey himself has done much to illuminate the processes involved and the impact which the relentless search of capital for new opportunities has had on geography. What his 1981 paper sought to argue is that this search for new opportunities in response to crises of over-accumulation also works itself at other levels, operating between countries within an international circuit of capital (=outer frontiers) as well as between core and peripheral regions within state economies (=inner frontiers). Just as crises of over-accumulation in existing areas and sectors of investment are solved by the migration of capital to new regions and new sectors within states, so also do we find capital seeking out investment opportunities in the growing network of colonies pieced together by European states from the sixteenth century onwards.

I have no disagreement with Harvey's interpretation of the instability behind capital formations or with his attempt to show how capital seeks out new areas and opportunities for investment, both within state systems and through the exploitation of the outer frontier areas provided by the forceful acquisition of colonies. However, the idea of a 'spatial fix' needs to be seen as a potential solution to a much wider range of inner conflicts than those associated with modern crises of capital over-accumulation. When we look at the historical record of frontier movements, many can be interpreted in terms of similar internal problems, with potential conflicts being released or resolved through an expansion into new peripheral areas. The history of Europe during, no less than after, the medieval period can be written out in terms of how its core societies interacted with their outer frontiers (Lewis 1958: 475–83; Chaunu 1979). Even during the eighth and ninth centuries, Charlemagne's Empire routinely used the challenge of its periphery to displace the rivalries and tensions of its core families (Smith 1996: 183–4). The Viking expansions of the ninth and tenth centuries can be seen as driven by internal population crises and conflicts over dynastic succession. As with the Carolingian Empire, these inner problems were resolved by being externalized, projected out over a space in which there was still, from a Norse viewpoint, a greater degree of 'unused freedom'.

We can apply the same sort of reasoning to what happened in Europe over the twelfth–fifteenth centuries. The extensive conquest and colonization of new areas that developed during this period was used both to release the tensions that had emerged in the older settled core areas of European feudalism by this point and to make critical changes that would have been far more difficult in core areas. Some of these problems derived from feudalism's character as a system of lordship. Initially, the land grants and jurisdictions that were given out by feudal kings in return for military services, and which served to give structure to the nascent feudal states of early medieval Europe, were not held on a hereditary basis. Whereas kin-based systems coped with population growth and competition for land and status through family settlements, feudal systems were not so responsive. To a degree, large territorial lordships could sub-divide themselves through a process of sub-infeudation. The essence of feudalism, though, was that grants of land and jurisdictions were held from the king and passed back to him on the death of the holder. Whilst those under the king had a vested interest in making their possession of land grants and jurisdictions hereditary, and allowing sons to share in them, this was not in the interest of state rulers. Though some lords eventually established hereditary claims, the rule of primogeniture meant that each generation produced a surplus of land-hungry baronial or knightly sons (Bartlett 1993: 47–51). We can see the effect of this in regard to Norman feudalism. Its westward and north-westward spread into new peripheral areas in Britain and Ireland over the eleventh and twelfth centuries, especially, provided expanded opportunities for its surplus sons, reducing what undoubtedly would have been a source of stress had the supply of land and opportunity remained finite.

Again, we can see this as a 'spatial fix', an external solution to the potential inner conflicts of feudal society.

Large-scale expansion along the eastern frontier of the Holy Roman Empire during the twelfth–fourteenth centuries provided a similar safety valve for feudal tensions in Germany. The push north-eastwards into Brandenburg and Prussia, eastwards into Poland and south-westwards into Austria and the middle Danube provided the younger sons of feudal lords, as well as many lesser tenants, with the sort of 'free-floating resources' that were not available in the older settled parts of Germany. It enabled Albrecht the Bear to build not just a new state but to 'establish a society de novo' (Thompson 1928: 519). Yet we must not overstate the extent to which Albrecht's frontier society was a different society as opposed to being simply a new extension to an existing society. In a searching review of medieval frontiers, Bartlett has suggested that frontiers like those fashioned by German expansion provided a huge reservoir of new opportunities for lordly families, especially their younger sons, and led to the creation of 'aristocratic diaspora'. In this respect, frontiers eased the problems of well-settled core areas, but could duplicate some of the institutional forms and structures found in core areas. However, Bartlett acknowledges that they had some potential for change. Indeed, in some cases, especially during the earlier phases of expansion, their 'freedom' actually led to the creation of purer feudal forms than existed in core areas (Bartlett 1989: 45). For this reason, Bartlett cautions against seeing frontier societies within a simple core–periphery model based on relations of dependency and exploitation (Bartlett 1993: 306–7).

Of course, the spread of lordship and knightly fees carried no force without peasants to work the soil. Not surprisingly, their spread into new and lightly settled areas was associated with a vigorous land colonization movement. We can see this colonization movement as part of a more widespread process that worked itself out at two scales. First, even within core areas, there existed areas of waste and less fertile soils into which the surplus produced by the rapidly growing population of the twelfth, thirteenth and early fourteenth centuries was able to spread. Though initially this filling out of settlement involved an extension of feudal tenures and relations, it soon became associated with the creation of wholly new forms of settlement based on cash-bearing rents, free tenures and new systems of husbandry. In effect, a geographically emplaced contrast arose between old-established feudal areas and new areas of settlement, with the latter being used to develop new relations of production that served to weaken feudalism from without (Dodgshon 1987: 240–86). We can match this contrast with the larger-scale contrast between the older settled parts of feudal Europe and the frontiers of settlement that were created around its outer edge. The settlement of these frontier areas involved a considerable movement of population from the more settled core areas 'into the continental peripheries' (Bartlett 1989: 23–47). As with the colonization of waste or peripheral areas within its core areas, this shift of colonists to the outer frontiers of feudal Europe involved the creation of settlements organized around freer, less feudally burdened, tenures (ibid.), though in parts of eastern Europe these freedoms were being eroded in the face of a more assertive lordship by 1500. However, at the outset, their settlement helps make the point that new spaces have long been used to facilitate change in a strategic way.

Spanish frontiers need a separate comment for there is a sense in which the Spanish frontier provides us with a hinge between the medieval and post-medieval world. The character of medieval Spain, at least from the eleventh century onwards, was greatly affected by the frontier spirit of the *reconquista*. The *reconquista* involved 'transforming disciplines' whose gradual frontier-like spread affected the nature of Spanish society deeply (MacKay 1977: 2). But once the *reconquista* had been accomplished, it left a society organized for a struggle that no longer existed. Seen in this context, the colonization and exploitation of the New World over the early sixteenth century served to subvert potentially destructive conflicts between the Spanish noblemen and the crown, with the former gaining hugely from the fresh opportunities opened up by the frontiers of the New World, and enabled the spirit of the frontier that had been developed fighting the Moors to be redirected (Wolf 1982: 113; Meinig 1986: 43–51).

As European states pushed the frontier of what was peripheral outwards on a global scale after 1492, so have such areas continued to provide 'free-floating resources' for them. They provided Europe with what has been called a 'ghost acreage' (Jones 1981: 70–84) in four ways. First, like the medieval frontiers skirting Europe itself, they provided a huge reservoir of land for colonization. Yet far from enabling migrants to remake Europe anew, in a wholly unrecognizable form, there is a case for arguing that the myriad cultural migrations involved, like all such migrations, gained not from the freedom which it offered to abandon the substance of their cultural practice or strategies but from the freedom it offered to shed the constraints that had previously ensnared them. Second, for some commentators, the large and sustained flow of European migrants to the Americas, Australasia and Africa across the sixteenth–nineteenth centuries saved parts of Europe from a Malthusian crisis and enabled increases in output to be translated into real gains in per capita growth. Third, indigenous populations in peripheral areas were coerced into providing a cheap labour force. The extensive use of slave labour especially, provided European core countries with access to forms of exploitation that might have been destabilizing if used in core societies. Fourth, the vast spaces drawn in through the colonial reach of European states opened up vast new areas of production. Combined with coercive labour systems, it enabled Europe's growing urban and industrial system to be resourced with new and cheap raw materials and for its labour force to be fed cheaply, and paid accordingly. Arguably, its own labour and production systems would not have permitted such a dramatic scale of growth, whatever their achievements in productivity. Indeed, a case has been made for seeing the windfall of colonial resources, both human and physical, to have been a more vital factor for Europe's economic growth than the value added by her own Industrial and Agricultural Revolutions (Blaut 1992: 36–53). Clearly, when seen in this wider context, the increasing tendency of investment capital from the nineteenth century onwards to seek a 'spatial fix' by exploiting opportunities overseas forms part of a more general strategy whereby core areas can export or displace their problems geographically. Without such strategy, or if the 'open-door' opportunities provided by peripheral areas had not been available, the history of Europe would surely have been different.

Core-driven change: upwards

A further means of core-driven expansion that works to displace the problems created by the accumulated inertia of core areas involves the exploitation of more hierarchical forms of organization, the spaces above. It works largely through the merger of, or co-operation between, like systems, forming larger, more integrated systems that require higher-order systems of administration. As current growth in the administration and policy of the European Union amply demonstrates, such a growth exploits new areas of choice and 'unused freedom'. By creating new spaces or levels of opportunity, such a growth bypasses problems of expended choice in established or mature systems. Infused with direction and purpose, it becomes a strategy of societal change, one that copes with inertia by circumventing it, or rather 'superventing' it. In character, it tends to involve groups of systems or organizations that are in some way competitive, with integration being a constructive way of controlling their rivalry just like hunter-gatherer bands who either fought each other or married each other. Certainly, Axelrod's 'TIT FOR TAT' lays out a feasible pathway of co-operation between players in a game of political rivalry, even where no central or co-ordinating authority is involved. Interestingly, Axelrod found that the time perspective was critical, for it was the durability or predictability of interaction, or knowledge about the history of their interaction, more than trust itself that fostered co-operation (Axelrod 1984: 182). As Axelrod also makes plain, governments and businesses were the forms most likely to seek co-operation, that is, those who might otherwise be rivals.

As regards political integration, we can see upward integration at work from the simplest polities onwards. Tribes and chiefdoms were particularly prone to forging complex alliance structures under an elected leader. Their problem was that for much of their history, these alliances were equally prone to collapsing back into their component parts. For Renfrew, the route out of this organizational sink was through the emergence of Early State Modules whose regular interaction and co-operation through exchange and military alliance led to state formation by integration (Renfrew 1979: 501; cf. Tilly 1975: 30). As noted in chapter 1, though, Renfrew did not see the crossing of the tribes–state boundary via the interim condition of chiefdoms as an irreversible step. Instead, he envisaged a phase during which conditions could produce either type of outcome and during which there may actually have been some switching between the two. Once in place, early state systems were subjected to two kinds of change. Some could display an eruptive imperialist growth, extending hegemony out over surrounding weaker states, chiefdoms and tribes. Others though, continued to grow through integration, as ruling dynasties worked to establish their greater kingship through alliance and claims of inheritance. Tilly talked about the reduction of European states from approximately 1,500 at the start of the medieval period to around 500 by 1500 as being the outcome of a ruthless competition fought out over organizational advantage as well as on the battlefield (ibid.: 30), and with co-operation being as much a factor as conquest. In one of the most recent reviews of how states develop, Watson has restated the extent to which such integration is endemic and the extent to which, as a process, it has remodelled political structures upwards, though he places very clear limitations on the desirability of continuing integration (Watson 1992: 319–22). Philbrick and Brown have provided a particularly systematic analysis of the geographical processes involved, one that draws out its seeming inevitability (Philbrick and Brown 1973: 61–90; see also, Isard 1975: 113–23; Cohen 1991: 551–80). What stands out from such analyses is how such integration allows for the ongoingness of established political structures. In a sense, it seems designed to build structure around inertia.

Trading systems have matched this capacity of political systems for merging and integrating into larger systems. During the early phases in the emergence of the world-trade system, trade followed the flag. By the nineteenth century though, trade was capable of establishing its own systems of integration. Thus, by the nineteenth century, Britain's empire can be seen as being as much informal as formal, the former being the integration achieved through trading domination alone as opposed to the advantage or domination gained through overt political control. Just as political integration produces a need for different layers of organization, so also has the growing scale, spread and connectivity of trade on a world scale generated the need for new layers of organization, thereby adding extra roles and opportunities *above* existing systems of trade. The way the world system has always been centred is a measure of this tendency for trade to become organized and hierarchical as it expanded. What began in centres like Venice, Antwerp and Amsterdam as the point at which maximum information about trading opportunities and a maximum range of goods were brought together developed into a more elaborate form as trade became as much a problem of organizing capital flows and investments, centres of information and price determination, as of creating gateways through which goods actually flowed. Over and above patterns of trade and production, these centres act as base-points for the system at large, helping to route capital in the direction of opportunity. As the narrow arteries of global trade already apparent by the sixteenth century were thickened and reticulated, the more burgeoning centres of control could seek wholly new opportunities by extending their activities upwards through innovations in organization and control as well as outwards by geographical extension. In the process, the channelling of both trade and its control led to huge savings in transaction costs (North 1981: 206).

Change from without and peripheral regions

The extensive discussion of core–periphery concepts and notions of internal colonialism has tended to portray peripheral spaces as areas of dependency, with conditions in them being driven by the needs of the core. Wallerstein's reference to underdevelopment being a structural condition, the necessary downside of development within a capitalist world system, comes to mind here. Such interpretations undoubtedly have some validity. Uneven development is an acknowledged by-product of capitalism in operation (Smith 1984: 87–104; Smith 1989: 142–63). However, we cannot ignore other critical aspects of the problem. Relative to core areas, peripheral and marginal areas also differ in ways that fit less comfortably into the structural or relational framework laid down by world-system theory or colonialism models.

Whilst all societies are inclined towards inertia and a consecutiveness of choice, core and peripheral areas or spaces differ as to the degree of inertia and consecutiveness involved. As the centres of co-ordination and consumption for societal systems, core areas are underpinned by far more mapped and bound information, more expended choice over how resources should be used. Furthermore, a high proportion of this information, its articulation in terms of symbolic, organizational and fixed capital forms, is orientated around the singular needs and norms of the system at large. For these reasons, such areas have a far larger proportion of resource and certainly far more in absolute terms committed to the mere maintenance or ongoingness of information, its inertia. By comparison, peripheral areas are likely to possess more 'unused freedom' over resources and information, to be less oppressed by the instrumental symbolism designed to ensure the continuity of the system at large. As with all forms of territorial organization, the density of decision- makers and weight of decision-making thins out towards peripheral and marginal areas simply because the ordination of power tends to be organized nodally, whether one is dealing here with political or trading systems. The simple fact that such areas are likely to be less developed and to have less information to process and store means that, relative to core areas, they are likely to have more flexibility even though they might be labelled as more 'tradition-bound' areas within the system as a whole. The key to this latter paradox is that tradition-bound areas tend to have not only less information to sustain, but also, to possess far more scope for reducing its redundancy, or patterning, without loss of meaning or information, thanks to the greater generalization and duplication of its roles and functions. When their latent flexibility is combined with the fact that, relative to cores, peripheral areas tend to have more potential hybridity and divergence, more niches within which difference can develop, we can understand why historical instances of runaway growth have regularly erupted out of such areas.

Yet as a number of commentators have noted when talking about how local elites in peripheral areas used their knowledge and links with core hegemonies to advantage, peripheral areas only rarely erupted in complete isolation. There was usually a symbiosis between the two. Just as core areas could exploit the physical and social distance of peripheral areas by exporting to them the kind of social devaluation of capital that might be deemed unacceptable closer to the core, so also could the latter develop part of their success out of the information, capital and technology introduced through their association with existing core areas (Gilpin 1981: 181; Meinig 1986: 373–4; Mann 1986: 187; Watson 1992: 128). In effect, the link between core and periphery acts as a ratchet, enabling some of the critical advantages or skills of existing core areas to be carried forward in combination with more indigenously derived changes. The extent to which peripheral areas could seize the advantage for themselves in this way is shown by the extent to which it actually happened. The history of early civilizations like those of China and the Middle East involved areas that initially were peripheral seizing the advantage. Examples of highly mobile pastoral societies from the heart of Asia seizing the political as well as the military initiative at the expense of settled societies around the Asian rim, from China and India to Arabia and Anatolia, abound (Chaudhuri 1990: 269–77). The same was true of Europe, with the location of core or leading-edge societies migrating across Europe (Mann 1986: 538–9). When we look at the history of the past century or so, these shifts have clearly not ceased. Hegel's notion of the lead 'civilizations', or leading-edge societies, shifting progressively from the Far East to Europe can now be rounded off into a completed circle or circumnavigation. The initial growth of America can be read as a former peripheral area combining skills from the core with the opportunities offered by vast new areas of choice and flexibility. In still more recent times, the eruptive emergence of the Pacific Rim as a centre of innovative and rapid economic growth adds to the list of emergent peripheries, though, like Europe and America before it, we can already detect strategic shifts within the broad area of the Pacific Rim.

Some concluding thoughts

The idea that beyond the individual instances of change, there may be processes or patterns about which we can generalize does not appeal to everyone.

Seeing it as an assumption about history that is shared by a number of 'pernicious postulates', Tilly has recently proclaimed that there is 'no such thing as social change in general' (Tilly 1984: 33). Even trying to shift the emphasis onto the question of where change takes place, as this book has done, does not escape such criticism. In a recent work, Harvey has argued that 'everyone who lives, acts and talks is implicated' in change (Harvey 1996: 106). The inevitable corollary of this is that change 'is everywhere' (ibid.: 105). In making this point, he was actually tilting at those who, like the present author, would see margins as spaces 'outside of the dominance of systems of determination' and 'the iron cage of circular and cumulative causation' and, therefore, as 'more revolutionary' (ibid.: 100). Tagging the idea as 'one of the least admirable traits' of recent political argument, Harvey directed his criticism at the work of writers like bell hooks who have defined peripheral or marginal spaces within modern urban societies as spaces of change. In fact, as I have tried to show, a great deal of discussion now exists demonstrating that some of the root concepts behind what hooks has to say are but a small part of a much larger argument. Furthermore, such ideas have been developed across a range of different historical contexts and by scholars from a range of different backgrounds. Indeed, Harvey's own work on capital and the 'spatial fix' can be seen as contributing to this wider argument.

In a sense, Harvey's dilemma mirrors the wider problem. On the one hand, there is a sense in which it is impossible to disagree with his view that some degree of change has always been inherent in the human condition. Yet on the other, any sort of familiarity with the human record of societal change also allows one to appreciate why Chaudhuri, drawing inspiration from Braudel, could talk of Asian societies having a 'perpetual memory' or 'replicative code' and of their structure as manifesting a 'temporal invariance' in the very long term that together foster variations on a theme rather than fundamentally new forms (Chaudhuri 1990: 10–11 and 375–7). To an extent, such differences can be accommodated by recognizing that societal change is richly dimensional in character, with each dimension having a different time constant. Braudel's structuring of the problem through different scales of analysis, each scale turning over at a different rate and, as a consequence, having a different inertial value, makes the same point (Braudel 1972: 14–17).

Another way of negotiating between such differences, though, is by recovering their geography. Harvey responds sceptically to the idea that peripheral spaces might be more capable of 'radical openness and revolutionary possibility' (Harvey 1996: 110). Yet such spaces, as Bhabha especially has argued, may have more hybridity and may tolerate, to use Allen's phrase, more 'hillclimbers'. They may also possess greater flexibility from being less co-ordinated by, or integrated into, a dominant system of values and thought. There is also a sense, though, in which what may really matter is their greater freedom of action rather than thought. Such a shift in emphasis is significant.

Even if we assume that there is an equal capability for thinking out new ideas amongst all societies, whether core or peripheral, old or new, it does not follow that all are equally capable or interested in responding to them. Indeed, there is a case for arguing that the question should be rephrased in terms of what constrains rather than what promotes change. At the heart of this point is the fact that despite attempts to characterize them as spaces of retardation, spaces left behind by the growth of core areas, peripheral spaces and their societies have been centres for the eruptive growth of new cores on a recurrent basis, at least when seen over a sufficiently long time span. In seeking to account for this, we need to invert how we see inertia. Far from being a dysfunction, inertia can be seen as a process fundamental to the ongoingness of any society. Instead of being the defining characteristic of marginal, or of traditional societies, there is a case for arguing that it is actually a more prominent and intrinsic feature of stable and mature societies, including the most complex, simply because of the greater volume of differentiated and specialized information that needs to be carried forward by them. In these circumstances, inertia serves to ensure stability through a consecutiveness of choice and its implied flow of accepted meaning, constraining change in the process.

Such inertia in the midst of complexity derives from a society's archaeology of past experience, that which it carries forward either as norms, codes and values or embodied in the symbolization, institutionalization and capitalization of the space through which it orders and identifies itself. It is this embeddedness to inertia, its geography or spatiality plain and simple, that helps configure the possibilities for change in the landscape. It makes the point, and does so forcibly, that change is not solely about perceived gains, but equally, about how it relates – through a calculation of potential risks, losses and devaluations – to what already is and its stored capital as a viable system of expended choice. This is why for some commentators, like Schrecker (1948: 250–72), any theory of how civilization develops must rest on an understanding of its freedom to respond to change, its stock of free-floating resources, not simply its capacity to imagine change. Needless to say, it is the geographical variations in this freedom that provide the basis for a geographically informed concept of change.

By way of a reply, it could be argued that if all societal change is contingent, or conjunctural, then it hardly matters whether the forces or institutional forms brought together within an episode of change are deep-rooted or shallow. All that matters is their configuration at the moment when they come together. Such reasoning, though, ignores the fact that most of the knowledge applied to living in the world at any one moment is, strictly speaking, prior knowledge, knowledge that is historically constituted. Further, this carry over of knowledge and its meaning does not amount to a *vis inertiae*, a dead force. Because of the way society creates new knowledge and new structures at the margins, literally and metaphorically, and because of the length of time required for the turnover of its more deeply embedded structures, as well as its more integrated structures, what is necessarily carried forward is a vital and active part of how societies conceive and interpret the present, part of what is still in the making. In other words, if change is always 'our next best move', then, as Vickers also reasoned, it does not mean that change can be reduced only to the moment in being and its prevailing conditions (Vickers 1994: 203). In so far as each 'decision margin' draws on a society's archaeology of past experience, it is necessarily affected by what is antecedent to it, both in terms of what is deep-rooted as well as what is proximate. Indeed, if human culture is accumulative, then we can expect the 'decision margin' to become burdened with an ever increasing amount of historically constituted knowledge *that is still open to use*. For this reason, what makes each successive 'moment in being' potentially different is the compounded knowledge that is drawn upon from what is past and how we choose to symbolize it. We may be 'mortgaged to structures accumulated along the way' (Sahlins 1977: 227), but these are precisely what give meaning to the 'moment in being' and what, through their ongoing and net accumulation at a whole variety of scales, and growing interaction, make each successive moment both meaningful and distinguishable. In short, there is a case for arguing that we live the past in the belief that we are really living the present. If we follow Bourdieu in believing that 'to change the world, one has to change the way of making the world, that is, the vision of the world' (Bourdieu 1990: 137), then we are compelled – necessarily so – to start with how we envision the past.

References

Abell, P. 1991 (ed.), *Rational Choice Theory*, Aldershot: Edward Elgar Publishing

Abu-Lughod, J. 1989, *Before European Hegemony. The World System AD 1250–1350*, New York: Oxford University Press

  1993, 'Discontinuities and persistence: one world system or a succession of world systems', in Frank and Gills (eds.), *The World System*, pp. 278–91

Adams, J. S. 1970, 'Residential structure of mid western cities', *Annals of the Association of American Geographers* 60: 37–63

Adams, R. M. 1966, *The Evolution of Urban Society*, Chicago: Aldine Publishing Co.

Agnew, J. 1987, *The United States in the World Economy. A Regional Geography*, Cambridge: Cambridge University Press

Alföldy, G. 1985, *The Social History of Rome*, trans. by D. Brand and F. Pollock, London: Croom Helm

Alland, A. 1975, 'Adaptation', *Annual Review of Anthropology* 4: 58–73

Allen, P. M. 1988a, 'Evolution: why the whole is greater than the sum of the parts', in W. Wolff, C-J. Soeder and F. R. Drepper (eds.), *Ecodynamics. Contributions to Theoretical Ecology*, Berlin: Springer-Verlag, pp. 2–30

  1988b, 'Evolution, innovation and economics', in G. Dosi, C. Freeman, R. Nelson, G. Silverberg and L. Soete (eds.), *Technical Change and Economic Theory*, London: Pinter Publishers, pp. 95–119

Appadurai, A. 1990, 'Disjuncture and difference in the global cultural economy', in M. Featherstone (ed.), *Global Culture. Nationalism, Globalization and Modernity*, London: Sage, pp. 295–310

Arrow, K. J. 1974, *The Limits of Organization*, New York: W. W. Norton

Arthur, W. B. 1988, 'Urban systems and historical path dependence', in J. H. Ausubel and R. Herman (eds.), *Cities and their Vital Systems*, Washington: National Academic Press, pp. 85–97

Ashby, W. R. 1968a, 'Principles of the self-organizing system', in Buckley (ed.), *Modern Systems Research*, Chicago: Aldine, pp. 108–18. First published in 1962

  1968b, 'Variety, constraint and the law of requisite variety', in Buckley (ed.), *Modern Research Systems*, pp. 129–36

  1968c, 'Regulation and control', in Buckley (ed.), *Modern Systems Research*, pp. 296–303

Aulin, A. 1986, 'Notes on the concept of self steering', in F. Geyer and J. van der

Zouwen (eds.), *Sociocybernetic Paradoxes. Observation, Control and Evolution of Self-Steering Systems*, London: Sage, pp. 100–18

Axelrod, R. 1984, *The Evolution of Cooperation*, New York: Basic Books

Bach, R. L. 1980, 'On the holism of a world-system perspective', in T. K. Hopkins and I. Wallerstein (eds.), *Processes of the World System*, vol. III, Political Economy of the World System Annuals, London: Sage

Bairoch, P. 1974, 'Geographical structure and trade balance of European foreign trade', *Jnl of European Economic History* 111: 557–608

Baker, A. R. H. 1979, 'Settlement pattern evolution and catastrophe theory: a comment', *Transactions, Institute of British Geographers*, new series 4: 435–7

1982, 'On ideology and historical geography', in A. R. H. Baker and M. Billinge (eds.), *Period and Place. Research Methods in Historical Geography*, Cambridge: Cambridge University Press, pp. 233–43

1984, 'Reflections on the relations of historical geography and the Annales school of history', in A. R. H. Baker and D. Gregory (eds.), *Explorations in Historical Geography. Interpretative Essays*, Cambridge: Cambridge University Press, pp. 1–27

1996, 'On the history and geography of historical geography', *Rekisdi Chirigaku* 38: 1–24

Barkow, J. H. Cosmides, L. and Tooby, J. 1992 (eds.), *The Adapted Mind. Evolutionary Psychology and the Generation of Culture*, Oxford: Oxford University Press

Barraclough, G. 1984 (ed.), *The Times Atlas of World History*, London: Guild Publishing

Bartlett, R. 1989, 'Colonial aristocracies of the high middle ages', in R. Bartlett and A. MacKay (eds.), *Medieval Frontier Societies*, Oxford: Clarendon Press, pp. 23–47

1993, *The Making of Europe. Conquest, Colonization and Cultural Change*, Harmondsworth: Penguin Books

Basu, K., Jones, E. L. and Schlicht, E. 1987, 'The growth and decay of custom: the role of the new institutional economics in economic history', *Explorations in Economic History* 24: 11–21

Bateson, G. 1972, *Steps Towards an Ecology of Mind. Collected Essays in Anthropology, Psychiatry, Evolution and Epistemology*, London: Intertext Books

1980, *Mind and Matter. A Necessary Unity*, London: Fontana

Berger, P. L. and Luckmann, T. 1966, *The Social Construction of Society*, Harmondsworth: Penguin Books

Berry, B. J. L. 1989, 'Comparative geography of the global economy: cultures, corporations, and the nation state', *Economic Geography* 65: 1–18

Bhabha, H. 1990, 'Interview with Bhabha on the third space', in J. Rutherford (ed.), *Identity, Community and Difference*, London: Lawrence & Wishart, 207–21

1994, *The Location of Culture*, London: Routledge

Blaut, J. 1992, 'Fourteen ninety-two', in J. M. Blaut *et al.*, *1492. The Debate on Colonialism, Eurocenterism and History*, Trenton: Africa World Press, pp. 1–63

Blute, M. 1979, 'Sociocultural evolutionarism: an untried theory', *Behavioural Science* 24: 46–59

Bock, R. E. 1963, 'Evolution, function and change', *American Sociological Review* 28: 229–37

Boudon, R. 1986, *Theories of Social Change. A Critical Appraisal*, Cambridge: Polity Press. First published in 1984

Boulding, K. E. 1956, 'Towards a general theory of growth', *General Systems* 1: 66–75
  1971, 'The economics of knowledge and the knowledge of economics', in D. M. Lamberton (ed.), *Economics of Information and Knowledge. Selected Readings*, Harmondsworth: Penguin Books, pp. 21–36
  1981, *Evolutionary Economics*, London: Sage

Bourdieu, P. 1977, *Outline of Theory as Practice*, Cambridge: Cambridge University Press
  1979, 'The Kabyle house or the world reversed', in P. Bourdieu, *Algeria 1960*, Cambridge: Cambridge University Press, pp. 133–53
  1990, *In Other Words. Essays Towards a Reflexive Sociology*, Stanford: Stanford University Press
  1992, 'Questions and answers', in P. Bourdieu and L. J. D. Wacquant (eds.), *An Invitation to Reflexive Sociology*, Cambridge: Polity Press, pp. 61–215

Bowen, M. 1981, *Empiricism and Geographical Thought*, Cambridge: Cambridge University Press

Braudel, F. 1972, 'History and the social sciences', in P. Burke (ed.), *Economy and Society in Early Modern Europe*, London: Routledge, pp. 11–42. First published in 1958
  1977, *Afterthoughts on Material Civilization and Capitalism*, Baltimore: Johns Hopkins University Press

Brenner, R. 1977, 'The origins of capitalist development: a critique of neo-Smithian Marxism', *New Left Review* 104: 25–92

Buckley, W. 1967, *Sociology and Modern Systems Theory*, Englewood Cliffs: Prentice Hall
  1968a (ed.), *Modern Systems Research for the Behavioral Scientist*, Chicago: Aldine Publishing Co.
  1968b, 'Society as a complex adaptive system', in Buckley (ed.), *Modern Systems Research*, pp. 490–513

Burns, R. I. 1989, 'The significance of the frontier in the middle ages', in R. Bartlett and A. MacKay (eds.), *Medieval Frontier Societies*, Oxford: Clarendon Press, pp. 307–30

Butlin, R. 1993, *Historical Geography. Through the Gates of Time and Space*, London: Edward Arnold

Butzer, K. 1980, 'Civilization: organisms or systems', *American Scientist* 68: 517–23

Campbell, B. M. S., Galloway, J. A., Keene, D. and Murphy, M. 1993, *A Medieval Capital and its Grain Supply. Agrarian Production and Distribution in the London Region c. 1300*, Historical Geography Research Series, no. 30

Carneiro, R. 1967, 'On the relationship between size of population and complexity of social organization', *Southwestern Jnl of Anthropology* 23: 234–43

Casseti, E. 1981, 'A catastrophe model of regional dynamics', *Annals of the Association of American Geographers* 71: 572–9
  1984, 'Peripheral growth in mature economies', *Economic Geography* 60: 122–31

Castells, M. 1989, *The Informational City. Information Technology, Economic Restructuring and the Urban-Regional Process*, Oxford: Basil Blackwell

Chandler, A. D. 1988, *The Essential Alfred Chandler. Essays Towards a Historical Theory of Big Business*, ed. T. McCraw, Boston: Harvard Business School Press

1990, *Scale and Scope. The Dynamics of Industrial Capitalism*, Cambridge, Mass.: Belknap Press

Chandler, A. D. and Redlich, F. 1988, 'Recent developments in American business administration and their conceptualization', in Chandler, *The Essential Alfred Chandler*, pp. 118–39

Chapman, S. D. 1970, 'Fixed capital formation in the British cotton industry, 1770–1815', *Economic History Review*, 2nd series 23: 235–66

Chaudhuri, K. N. 1990, *Asia Before Europe. Economy and Civilization from the Rise of Islam to 1750*, Cambridge: Cambridge University Press

Chaunu, P. 1979, *European Expansion in the Later Middle Ages*, trans. by K. Bertram, Amsterdam: North Holland. First published in 1969

Chi, C. 1965, 'History governed by key economic areas', in Meskill (ed.), *The Pattern of Chinese History*, pp. 65–8

Chomsky, N. 1976, *Reflections on Language*, London: Fontana

Clanchy, M. T. 1970, 'Remembering the past and the good old law', *History* 55: 165–76

Clarke, D. L. 1968, *Analytical Archaeology*, London: Methuen

Claval, P. 1989, 'New interpretations of the French Revolution and their geographical significance', *Jnl of Historical Geography* 15: 260–8

Cohen, S. B. 1991, 'Global geopolitical change in the post-cold war era', *Annals of the Association of American Geographers* 81: 551–80

Conzen, M. 1990, 'Ethnicity on the land', in M. Conzen (ed.), *The Making of the American Landscape*, New York: Harper Collins, pp. 221–48

Corrigan, P. and Sayer, D. 1985, *The Great Arch. English State Formation as a Cultural Revolution*, Oxford: Basil Blackwell

Cosmides, L., Tooby, J. and Barkow, J. H. 1992, 'Introduction: evolutionary psychology and conceptual integration', in Barkow, Cosmides and Tooby (eds.), *The Adapted Mind*, pp. 3–15

Cottrell, P. L. 1980, *Industrial Finance 1830–1914. The Finance and Organization of English Manufacturing Industry*, London: Methuen

Coulburn, R. 1954, 'Structure and process in the rise and fall of civilized society', *Comparative Studies in Society and History* 8: 404–31

Crang, P. 1994, 'On the heritage trail: maps of and journeys to olde Englande', *Environment and Planning, D: Society and Space* 12: 341–55

Crawford, H. 1991, *Sumer and the Sumerians*, Cambridge: Cambridge University Press

Curry, L. 1964, 'The random spatial economy: an exploration in settlement theory', *Annals of the Association of American Geographers* 54: 138–46

1972, 'Chance and landscape', in P. W. English and R. C. Mayfield (eds.), *Man, Space and Environment*, New York: Oxford University Press, pp. 611–23. First published in 1966

Darby, H. C. 1962, 'Historical geography', in H. P. R. Finberg (ed.), *Approaches to History*, London: Routledge & Kegan Paul

1977, *Domesday England*, Cambridge: Cambridge University Press

Davis, L. E. 1971, 'Capital mobility and American growth', in R. W. Fogel and S. L. Engerman (eds.), *The Reinterpretation of American Economic History*, New York: Harper & Row: pp. 285–300

Dear, M. and Wolch, J. 1989, 'How territory shapes social life', in J. Wolch and M. Dear (eds.), *The Power of Geography. How Territory Shapes Social Life*, London: Unwin Hyman, pp. 3–18

Denevan, W. M. 1983, 'Adaptation, variation and cultural geography', *Professional Geographer* 35: 399–407

Deutsch, K. 1963, *The Nerves of Government*, New York: Free Press

Dodgshon, R. A. 1985, 'Symbolic classification and the making of early Celtic landscape', *Cosmos* 1: 61–83

1987, *The European Past. Social Evolution and Spatial Order*, London: Macmillan

1988, 'The Scottish farming township as metaphor', in L. Leneman (ed.), *Perspectives in Scottish Social History*, Aberdeen: Aberdeen University Press, pp. 69–82

1991, 'The changing evaluation of space 1500–1914', in R. A. Dodgshon and R. A. Butlin, *An Historical Geography of England and Wales*, London: Academic Press, pp. 255–83

1993, 'The modern world-system: a critique of its inner dynamics', in H-J. Nitz (ed.), *The Early Modern World System in Geographical Perspective*, Stuttgart: Verlag, pp. 26–41

Domar, E. D. 1961, 'The capital–output ratio in the United States: its variation and stability', in F. A. Lutz and D. C. Hague (eds.), *The Theory of Capital*, London: Macmillan, pp. 95–117

Donald, M. 1991, *Origins of the Modern Mind. Three Stages in the Evolution of Culture and Cognition*, Cambridge: Harvard University Press

Doyle, M. W. 1986, *Empires*, Ithaca: Cornell University Press

Driver, F. 1992, 'Geography and power: the work of Michel Foucault', in P. Burke (ed.), *Critical Essays on Michel Foucault*, Aldershot: Scolar, pp. 147–56

Duncan, J. 1990, *The City as Text. The Politics of Landscape Interpretation in the Kandyan Kingdom*, Cambridge: Cambridge University Press

Edelstein, M. 1971, 'Rigidity and bias in the British capital markets, 1870–1913', in D. N. McCloskey (ed.), *Essays on the Mature Economy. Britain After 1840*, London: Methuen, pp. 83–111

Eisenstadt, S. N. 1963, *The Political Systems of Empires*, New York: Free Press of Glencoe

1964, 'Institutionalization and change', *American Sociological Review* 29: 235–47

Ekholm, K. 1980, 'On the limitations of civilizations: the structure and dynamics of global systems', *Dialectical Anthropology* 5: 155–66

1981, 'On the structure and dynamics of global systems', in J. S. Kahn and J. R. Llobera (eds.), *The Anthropology of Pre-capitalist Societies*, London: Macmillan, pp. 241–61

Ekholm, K. and Friedman, J. 1993, '"Capital", imperialism and exploitation in ancient world systems', in Frank and Gills (eds.), *The World System*, pp. 59–80

Elbraum, B. and Lazonick, W 1986 (eds.), *The Decline of the British Economy: An Institutional Perspective*, Oxford: Clarendon Press

Eliade, M. 1971, *The Myth of the Eternal Return or, Cosmos and History*, Princeton: Princeton University Press

1973, *Australian Religions. An Introduction*, Ithaca: Cornell University Press

Elton, G. R. 1953, *The Tudor Revolution in Government. Administrative Changes in the Reign of Henry VIII*, Cambridge: Cambridge University Press

1960, *The Tudor Constitution. Documents and Commentary*, Cambridge: Cambridge University Press

Elvin, M. 1973, *Pattern of the Chinese Past*, London: Eyre Methuen

Fabian, J. 1983, *Time and the Other. How Anthropology Makes its Object*, New York: Columbia University Press

Falkus, M. E. 1977, 'The development of municipal trading in the nineteenth century', *Business History* 19: 134–56

Featherstone, M. 1995, *Undoing Culture. Globalization, Postmodernism and Identity*, London: Sage

Feinstein, C. H. 1981, 'Capital accumulation and the industrial revolution', in R. Floud and D. McCloskey (eds.), *The Economic History of Britain since 1700, vol. I, 1700–1860*, Cambridge: Cambridge University Press, pp. 128–42

Feinstein, C. H., and Pollard, S. 1989, *Studies in Capital Formation in the United Kingdom 1750–1920*, Oxford: Clarendon Press

Flannery, K. V. 1976, 'The cultural evolution of civilizations', in P. J. Richerson and J. McEvoy III (eds.), *Human Ecology. An Environmental Approach*, North Scituate, Mass: Duxbury, pp. 96–118

Florida, R. 1996, 'Regional creative destruction: production organization, globalization and the economic transformation of the Midwest', *Economic Geography* 72: 314–34

Foucault, M. 1974, *The Order of Things. An Archaeology of the Human Sciences*, London: Routledge

Fowler, P. J. 1987, 'The contemporary past', in J. M. Wagstaff (ed.), *Landscape and Culture. Geographical and Archeological Perspectives*, Oxford: Basil Blackwell, pp. 173–91

Fox, W. 1971, *History in Geographic Perspective. The Other France*, New York: W. W. Norton & Co.

Frank, A. G. and Gills, B. K. 1993a (eds.), *The World System. Five Hundred Years or Five Thousand?*, London: Routledge
    1993, 'The 5000-year old world system', in Frank and Gills (eds.), *The World System*, pp. 3–55

Fraser, D. 1982, 'Introduction: municipal reform in historical perspective', in D. Fraser (ed.), *Municipal Reform and Industrial City*, Leicester: Leicester University Press, pp. 1–14

Friedman, J. and Rowlands, M. J. 1977, 'Notes towards an epigenetic model of the evolution of civilization', in J. Friedman and M. J. Rowlands (eds.), *The Evolution of Social Systems*, London: Duckworth, pp. 201–76

Fry, G. K. 1979, *The Growth of Government*, London: Cass

Gallman, R. E. 1966, 'Gross national product in the United States 1834–1909', in *Output, Employment and Productivity in the United States after 1800*, National Bureau of Economic Research (New York), New York: Columbia University Press, pp. 3–76

Garnsey, P. D. A. 1978, 'Rome's African empire under the principate', in P. D. A. Garnsey and C. R. Whittaker (eds.), *Imperialism in the Ancient World*, Cambridge: Cambridge University Press, pp. 223–54
    1983, 'Grain for Rome', in P. D. A. Garnsey, K. Hopkins and C. R. Whittaker (eds.), *Trade in the Ancient Economy*, Berkeley: University of California Press, pp. 118–30

Garnsey, P. D. A. and Saller, R. 1987, *The Roman Empire. Economy, Society and Culture*, London: Duckworth

Gellner, E. 1964, *Thought and Change*, London: Weidenfeld & Nicolson

1983, *Nations and Nationalism*, Ithaca: Cornell University Press

1988, 'Introduction', in J. A. Hall and M. Mann (eds.), *Europe and the Rise of Capitalism*, Oxford: Basil Blackwell, pp. 1–5

Gertler, M. S. 1984, 'The dynamics of regional capital accumulation', *Economic Geography* 60: 150–174

1988, 'The limits to flexibility: comments on the post-Fordist version of production and its geography', *Transactions of the Institute of British Geographers*, new series 13: 419–32

Gibson, M. 1973, 'Population shift and the rise of Mesopotamian civilization', in C. Renfrew (ed.), *The Explanation of Culture Change. Models in Prehistory*, London: Duckworth, pp. 447–63

Giddens, A. 1981, *Contemporary Critique of Historical Materialism*, vol. I, Power, Property and the State, London: Macmillan

1984, *The Constitution of Society*, Cambridge: Polity Press

Gills, B. K. 1993, 'Hegemonic transitions in the world system', in Frank and Gills (eds.), *The World System*, pp. 115–40

Gills, B. K. and Frank, A. G. 1993a, 'The cumulation of accumulation', in Frank and Gills (eds.), *The World System*, pp. 81–114

1993b, 'World system cycles, crises and hegemonic shifts, 1700 BC to AD 1700', in Frank and Gills (eds.), *The World System*, pp. 143–99

Gilpin, R. 1981, *War and Change in World Politics*, Cambridge: Cambridge University Press

Goldthwaite, R. A. 1980, *The Building of Renaissance Florence. An Economic and Social History*, Baltimore: Johns Hopkins University Press

Goodwin, M. and Painter, J. 1996, 'Local governance, the crises of Fordism and the changing geographies of regulation', *Transactions of the Institute of British Geographers*, new series 21: 635–48

Goody, J. 1986, *The Logic of Writing and the Organization of Society*, Cambridge: Cambridge University Press

Gould, P. 1972, 'Pedagogic review', *Annals of the Association of American Geographers* 62: 689–700

Gould, S. J. 1996, *Life's Grandeur. The Spread of Excellence from Plato to Darwin*, London: J. Cape

Grassby, R. 1970, 'English merchant capitalism in the late seventeenth century: the composition of business fortunes', *Past and Present* 46: 87–107

Gregory, D. 1980, 'The ideology of control: systems theory and geography', *Tijdschrift voor Economische en Sociale Geografie* 71: 327–42

1981, 'Human agency and human geography', *Transactions of the Institute of British Geographers*, new series 6: 1–18

1989, 'The crisis of modernity? Human geography and critical social theory', in R. Peet and N. Thrift (eds.), *New Models in Geography*, vol. II, London: Unwin Hyman, pp. 348–85

1990a, 'Grand maps of history: structuration theory and social change', in J. Clark, C. Modgill and S. Modgill (eds.), *Anthony Giddens. Consensus and Controversy*, Brighton: Falmer Press, pp. 217–33

1990b, 'A new and differing face in many places: three geographies of industrialization', in R. A. Dodgshon and R. A. Butlin (eds.), *An Historical Geography of England and Wales*, London: Academic Press, pp. 351–99

1991, 'Interventions in the historical geography of modernity: social theory, spatiality and the politics of representation', *Geografiska Annaler* 73B (1991): 17–44

1994, *Geographical Imaginations*, Oxford: Basil Blackwell

Gregory, D. and Urry, J. 1985 (eds.), *Social Relations and Spatial Structure*, London: Macmillan

Gregson, N. 1986, 'On duality and dualism: the case of structuration and time geography', *Progress in Human Geography* 10: 184–205

Griaule, M. and Dieterlen, G. 1954, 'The Dogon', in D. Forde (ed.), *African Worlds*, London: RIIA, pp. 83–110

Griffin, T. J. 1976, 'Revised estimates of the consumption and stock of fixed capital', *Economic Trends*, no. 264, London: HMSO

1979, 'The stock of fixed assets in the United Kingdom: how to make best use of the statistics', in K. D. Patterson and K. Schott (eds.), *The Measurement of Capital. Theory and Practice*, London: Macmillan, pp. 99–132

Habermas, J. 1979, *Communication and the Evolution of Society*, London: Heinemann Education

1984, *The Theory of Communicative Action, vol. I, Reason and the Rationalization of Society*, trans. by T. McCarthy, London: Heinemann Education

1987, *The Theory of Communicative Action, vol. II, The Critique of Functionalist Reason*, trans. by T. McCarthy, Cambridge: Polity Press

1990, *The Philosophical Discourse of Modernity*, Cambridge: Polity Press. Originally published 1987

Haeckel, S. H. and Nolan, R. L. 1993–4, 'The role of technology in an information age: transforming symbols into action', in *The Knowledge Economy. The Nature of Information in the 21st Century*, Nashville: Institute for Information Studies, pp. 1–24

Hall, J. A. 1986, *Powers and Liberties. The Causes and Consequences of the Rise of the West*, Harmondsworth: Penguin Books. First published in 1985

1988, 'States and societies: the miracle in comparative perspective', in J. Baechler, J. A. Hall and M. Mann (eds.), *Europe and the Rise of Capitalism*, Oxford: Basil Blackwell, pp. 20–38

Hall, P. and Preston, P. 1988, *The Carrier Wave. New Information Technology and the Geography of Innovation*, London: Unwin Hyman

Hall, S. 1991, 'The local and the global: globalization and ethnicity', in A. D. King (ed.), *Culture, Globalization and the World System. Contemporary Conditions for the Representation of Identity*, London: Macmillan, pp. 19–39

Hallpike, C. R. 1986, *The Principles of Social Evolution*, Oxford: Clarendon Press

Harley, C. K. 1988, 'Ocean freight rates and productivity, 1740–1913: the primacy of mechanical inventions reaffirmed', *Jnl of Economic History* 48: 851–76

Harris, C. 1978, 'The historical mind and the practice of geography', in D. Ley and M. Samuels (eds.), *Humanistic Geography. Prospects and Problems*, Chicago: Maroufa Press, pp. 123–37

1991, 'Power, modernity and historical geography', *Annals of the Association of American Geographers* 81: 671–83

Harris, M. 1979, *Cultural Materialism. The Struggle for a Social Science of Culture*, New York: Vintage Books

Harvey, D. 1978, 'Urbanization under capitalism: a framework for analysis', *International Jnl of Urban and Regional Research* 2: 101–31

1981, 'The spatial fix – Hegel, von Thünen, and Marx', *Antipode* 13: 1–12

1982, *The Limits of Capital*, Oxford: Basil Blackwell

1985, *The Urbanization of Capital*, Oxford: Basil Blackwell

1989, *The Condition of Postmodernity. An Enquiry into the Origins of Cultural Change*, Oxford: Basil Blackwell

1990, 'Between space and time: reflections on the geographical imagination', *Annals of the Association of American Geographers* 80: 418–34

1996, *Justice, Nature and the Geography of Difference*, Oxford: Basil Blackwell

Hawke, G. R. and Reed, M. C. 1969, 'Railway capital in the United Kingdom in the nineteenth century', *Economic History Review*, 2nd series 22: 269–86

Hayden, D. 1995, *The Power of Place. Urban Landscapes as Public History*, Cambridge, Mass.: MIT Press

Head, L. 1993, 'Unearthing prehistoric cultural landscapes: a view from Australia', *Transactions of the Institute of British Geographers*, new series 18: 481–99

Hegel, G. W. F. 1821, *Philosophy of Right*, 1896 edn, trans. by S. W. Dyde, London: G. Bell & Sons

Heitland, W. E. 1962, 'The Roman fate', in Kagan (ed.), *Decline and Fall of the Roman Empire*, pp. 57–70

Heper, M. 1985, 'The state and public bureaucracy: a comparative and historical perspective', *Comparative Studies in Society and History* 27: 86–110

Herlihy, D. 1967, *Medieval and Renaissance Pistoia. The Social History of an Italian Town, 1200–1430*, New Haven: Yale University Press

Hicks, J. 1969, *A Theory of Economic History*, Oxford: Clarendon Press

1973, *Capital and Time*, Oxford: Clarendon Press

Hill, C. 1969, *Reformation to Industrial Revolution*, Harmondsworth: Penguin Books

Hindness, B. and Hirst, P. Q. 1975, *Pre-capitalist Modes of Production*, London: Routledge & Kegan Paul

Hirschman, A. O. 1955, *The Strategy of Economic Development*, New Haven: Yale University Press

1977, *The Passions and the Interests. Political Arguments for Capitalism before its Triumph*, Princeton: Princeton University Press

Hocart, H. M. 1970, *Kings and Councillors. An Essay in the Comparative Anatomy of Human Society*, Chicago: Chicago University Press. First published in 1956

Hodder, I. 1990, *The Domestication of Europe*, Oxford: Basil Blackwell

Hoffman, W. G. 1961, 'Long-term growth and capital formation in Germany', in F. A. Lutz and D. C. Hague (eds.), *The Theory of Capital*, London: Macmillan, pp. 118–40

hooks, b. 1991, *Yearning. Race, Gender, and Cultural Politics*, London: Turnaround

Hopkins, K. 1980, 'Taxes and trade in the Roman empire (200 BC–AD 400)', *Jnl of Roman Studies* 70: 101–25

Hugill, P. J. 1993, *World Trade since 1431. Geography, Technology and Capitalism*, Baltimore: Johns Hopkins University Press

Ingham, G. 1984, *Capitalism Divided? The City and Industry in British Social Development*, London: Macmillan

Innis, H. A. 1972, *Empire and Communications*, rev. edn ed. M. Q. Innis, Toronto: University of Toronto Press. First published in 1950

Isard, W. 1975, 'Notes on an evolutionary theoretic approach to world organization', *Papers of the Peace Science Society (International)* 24: 113–23

   1976, 'On regional science models: parallels from biological science', *Papers of the Regional Science Association* 36: 77–85

   1977, 'Strategic elements of a theory of major structural change', *Papers of the Regional Science Association* 38: 1–14

Isard, W. and Liossatos, P. 1977, 'Space–time development and a general transfer principle', *Papers of the Regional Science Association* 30: 17–37

Ives, B. and Jarvenpaa, S. L. 1993–4, 'Competing with information: empowering knowledge networks with information technology', in *The Knowledge Economy*, Nashville: Institute for Information Studies, 53–87

Jameson, F. 1991, *Postmodernism, or the Cultural Logic of Late Capitalism*, London: Verso

Johnson, B. L. C. 1951–2, 'The Foley partnership: the iron industry at the end of the charcoal era', *Economic History Review* 4: 322–40

Johnson, C. 1964, *Revolution and the Social System*, Stanford: Stanford University Press

   1966, *Revolutionary Change*, Boston: Little, Brown & Co.

Johnson, G. A. 1982, 'Organizational structure and scalar stress', in C. Renfrew, M. J. Rowlands and B. A. Segraves (eds.), *Theory and Explanaton in Archaeology*, London: Academic Press, pp. 389–421

Jones, A. H. M. 1964, *The Later Roman Empire AD 284–602*, Oxford: Basil Blackwell
   1966, *The Decline of the Ancient World*, London: Longman

Jones, E. L. 1974, 'Industrial capital and landed investment: the Arkwrights in Herefordshire, 1809–43', in E. L. Jones, *Agriculture and the Industrial Revolution*, Oxford: Basil Blackwell, pp. 160–83. First published in 1967

   1981, *The European Miracle. Environments, Economies and Geopolitics in the History of Europe and Asia*, Cambridge: Cambridge University Press

   1988, *Growth Recurring. Economic Change in World History*, Oxford: Clarendon Press

Jones, E. L. and Falkus, M. E. 1979, 'Urban improvement and the English economy in the seventeenth and eighteenth centuries', *Research in Economic History* 4: 193–233

Jones, E. L., Porter, S. and Turner, M. 1984, *A Gazeteer of English Urban Fire Disasters 1500–1900*, Historical Geography Research Series, no. 13

Kagan, D. 1962 (ed.), *Decline and Fall of the Roman Empire. Why Did it Collapse?*, Boston: D. C. Heath & Co.

Kaufman, H. 1971, *The Limits of Organizational Change*, Alabama: University of Alabama Press

   1975, 'The natural history of human organizations', *Administration and Society* 8: 131–49

1976, *Are Government Organizations Immortal?*, Washington: the Brooking Institution

Keith-Lucas, K. 1980, *The Unreformed Local Government System*, London: Croom Helm

Kuznets, S. 1946, *National Income. A Summary of Findings*, New York: National Bureau of Economic Research

1955, 'International differences in capital formation and financing', in *Capital Formation and Economic Growth*, National Bureau of Economic Research (New York), Princeton: Princeton University Press, pp. 19–106

1971, 'Notes on the pattern of US economic growth', in R. W. Fogel and S. L. Engerman (eds.), *The Reinterpretation of American Economic History*, New York: Harper & Row, pp. 17–24

Langton J. and Höppe, G. 1983, *Town and Country in the Development of Early Modern Western Europe*, Historical Geography Study Group, no. 11

1995, *Peasantry to Capitalism. Western Östergötland in the Nineteenth Century*, Cambridge: Cambridge University Press

Larsen, M. T. 1988, 'Introduction: literacy and social complexity', in J. Gledhill, B. Bender, and M. T. Larsen (eds.), *State and Society. The Emergence of Social Hierarchy and Political Centralization*, London: Unwin Hyman, pp. 173–91

Lash, S. and Urry, J. 1994, *Economies of Signs and Symbols*, London: Sage

Laszlo, E. 1986, 'Systems and societies: the basic cybernetics of social evolution', in F. Geyer and J. van der Zouwen (eds.), *Sociocybernetic Paradoxes. Observation and Control of Self-Steering Systems*, London: Sage, pp. 145–71

Lattimore, O. 1940, *Inner Asian Frontiers of China*, American Geographical Society, research series no. 21, New York: American Geographical Society

1962, *Studies in Frontier History. Collected Papers 1928–58*, Oxford: Oxford University Press

Lawton, R. and Pooley, C. 1992, *Britain 1740–1950. An Historical Geography*, London: Longman

Leach, E. 1976, *Culture and Communication. The Logic by which Symbols are Communicated*, Cambridge: Cambridge University Press

Lee, J. S. 1965, 'The periodic recurrence of internecine wars in China', in Meskill (ed.), *Pattern of Chinese History*, pp. 25–30

Lefebvre, L. 1991, *The Production of Space*, trans. by D. Nicholson-Smith, Oxford: Basil Blackwell

Lévi-Strauss, C. 1966, *The Savage Mind*, London: Weidenfeld & Nicolson. First published in 1962

1968, *Structural Anthropology*, trans. by C. Jacobson, Harmondsworth: Allen Lane

1969, *The Elementary Structures of Kinship*, 2nd edn, London: Eyre and Spottiswoode. First published in 1949

Levick, B. 1985, *The Government of the Roman Empire. A Sourcebook*, London: Croom Helm

Lévy-Leboyer, M. 1978, 'Capital investment and economic growth in France, 1820–1930', in P. Mathias and M. M. Postan (eds.), *The Cambridge Economic History of Europe, vol. VII, The Industrial Economies: Capital, Labour and Enterprise, part 1, Britain, France, Germany and Scandinavia*, Cambridge: Cambridge University Press, pp. 231–95

Lewis, A. R. 1958, 'The closing of the medieval frontier', *Speculum* 33: 475–83

Lewit, T. 1991, *Agricultural Production in the Roman Economy, AD 200–400*, International Series, no. 568, Oxford: British Archaeological Reports

Ley, D. 1977, 'Social geography and the take-it-for-granted world', *Transactions of the Institute of British Geographers*, new series 2: 498–512

  1987, 'Styles of the times: liberal and neo-conservative landscapes in inner Vancouver, 1968–1986', *Jnl of Historical Geography* 13: 40–56

Lloyd, C. 1993, *The Structures of History*, Oxford: Basil Blackwell

Lowe, A. 1955, 'Structural analysis of real capital formation', in *Capital Formation and Economic Growth*, National Bureau of Economic Research (New York), Princeton: Princeton University Press, pp. 581–634

Lowenthal, D. 1975, 'Past time, present place: landscape and memory', *Geographical Review* 65: 1–36

  1976, 'The place of the past in the American landscape', in Lowenthal and Bowden (eds.), *Geographies of the Mind*, pp. 89–117

  1979, 'Age and artifact: dilemmas of appreciation', in D. W. Meinig (ed.), *The Interpretation of Ordinary Landscapes. Geographical Essays*, New York: Oxford University Press, pp. 103–28

  1985, *The Past is a Foreign Country*, Cambridge: Cambridge University Press

Lowenthal, D. and Bowden, M. J. (eds.) 1976, *Geographies of the Mind. Essays in Historical Geography in Honour of John Kirtland Wright*, New York: Oxford University Press

Lubenow, W. C. 1971, *The Politics of Government Growth. Early Victorian Attitudes Toward State Intervention 1833–1848*, Newton Abbot: David and Charles

Lyle, E. 1993, 'Internal–external memory', in S. Flood (ed.), *Mapping Invisible Worlds*, Cosmos, no. 9, Edinburgh: Edinburgh University Press, pp. 63–74

Lyotard, J-F. 1984, *The Post-Modern Condition. A Report on Knowledge*, Manchester: Manchester University Press. First published in 1979

MacDonagh, O. 1977, *Early Victorian Government, 1830–1870*, London: Weidenfeld & Nicolson

MacKay, A. 1977, *Spain in the Middle Ages. From Frontier to Empire 1000–1500*, London: Macmillan

Maisels, C. K. 1990, *The Emergence of Civilization. From Hunting and Gathering to Agriculture, Cities, and the State in the Near East*, London: Routledge

Mann, M. 1986, *The Sources of Social power, vol. I, A History of Power from the Beginning to AD 1760*, Cambridge: Cambridge University Press

Markusen, A. 1996, 'Sticky places in slippery space: a typology of industrial districts', *Economic Geography* 72: 293–334

Massey, D. 1984, *Spatial Divisions of Labour. Social Structures and the Geography of Production*, London: Macmillan

  1992, 'Politics and space/time', *New Left Review* 196: 65–84

Meinig, D. W. 1979, 'Symbolic landscapes: some idealizations of American communities', in D. W. Meinig (ed.), *The Interpretation of Ordinary Landscapes. Geographical Essays*, New York: Oxford University Press, pp. 164–92

  1986, *The Shaping of America. A Geographical Perspective on 500 Years of History, vol. I, Atlantic America 1492–1800*, New Haven: Yale University Press

  1989, 'A geographical transect of the Atlantic world, ca. 1759', in E. D. Genovese and L. Hochberg (eds.), *Geographic Perspectives in History*, Oxford: Basil Blackwell, pp. 185–204

1993, *The Shaping of America. A Geographical Perspective on 500 Years of History*, vol. *II, Continental America 1800–1867*, New Haven: Yale University Press

Merrifield, A. 1993, 'Place and space: a Lefebvrian reconciliation', *Transactions of the Institute of British Geographers*, new series 18: 515–31

Meskill, J. 1965 (ed.), *The Pattern of Chinese History. Cycles, Development or Stagnation?*, Boston: D. C. Heath

Mikesell, M. W. 1976, 'The rise and decline of "sequence occupance": a chapter in the history of American geography', in Lowenthal and Bowden (eds.), *Geographies of the Mind*, pp. 148–69

Modelski, G. 1983, 'Long cycles of world leadership', in W. R. Thompson (ed.), *Contending Approaches to World System Analysis*, Beverly Hills: Sage, pp. 115–39

Mokyr, R. J. 1990, *The Lever of Riches. Technological Creativity and Economic Progress*, Oxford: Oxford University Press

Moore, W. E. 1967, *Order and Change*, New York: Wiley

Myer, P. J. 1993–4, 'Foreword', in *The Knowledge Economy. The Nature of Information in the 21st Century*, Nashville: Institute for Information Studies, pp. v–viii

Myrdal, G. 1957, *Economic Theory and Under-developed Regions*, London: Duckworth

Nash, D. 1987, 'Imperial expansion under the Roman republic', in M. Rowlands, M. Larsen and K. Kristiansen (eds.), *Centre and Periphery in the Ancient World*, Cambridge: Cambridge University Press, pp. 87–103

Nebes, R. D. 1977, 'Man's so-called minor hemisphere', in M. C. Wittrock (ed.), *The Human Brain*, Englewood Cliffs: Prentice Hall, pp. 97–106

Needham, R. 1969, *Symbolic Classification*, Santa Monica: Goodyear Perspectives in Anthropology

Nelson, R. and Winter, S. 1982, *An Evolutionary Theory of Economic Progress*, Cambridge, Mass.: Belknap Press

Newson, L. 1976, 'Cultural evolution: a basic concept for human and historical geography', *Jnl of Historical Geography* 2: 239–55

Nisbet, R. 1969, *Social Change and History. Aspects of the Western Theory of Development*, Oxford: Oxford University Press

North, D. C. 1981, *Structure and Change in Economic History*, New York: W. W. Norton & Co.

Oates, J. 1977, 'Mesopotamian social organization: archaeological and philological evidence', in J. Friedman and M. J. Rowlands (eds.), *The Evolution of Social Systems*, London: Duckworth, pp. 457–85

O'Brien, P. 1982, 'European economic development: the contribution of the periphery', *Economic History Review*, 2nd series 35: 1–18

O'Driscoll, G. P. and Rizzo, M. J. 1985, *The Economics of Time and Ignorance*, Oxford: Basil Blackwell

Offer, A. 1981, *Property and Politics 1870–1914. Landownership, Law and Urban Development in England*, Cambridge: Cambridge University Press

Olson, M. 1982, *The Rise and Decline of Nations*, New Haven: Yale University Press

Olsson, G. 1982, '-/-', in P. Gould and G. Olsson (eds.), *A Search for Common Ground*, London: Pion Ltd, pp. 223–31

Ong, W. J. 1977, *Interfaces of the Word. Studies in the Evolution of Consciousness and Culture*, Ithaca: Cornell University Press

1982, *Orality and Literacy. The Technologizing of the Word*, London: Methuen

Parris, H. 1982, *Constitutional Bureaucracy*, London: Allen & Unwin

Parsons, T. 1964, 'Evolutionary universals in society', *American Sociological Review* 29: 339–57

1966, *Societies. Evolutionary and Comparative Perspectives*, Englewood Cliffs: Prentice-Hall Inc.

1977, *The Evolution of Societies*, ed. J. Toby, Englewood Cliffs: Prentice Hall. First published in 1966 as *The System of Modern Societies*, Englewood Cliffs: Prentice Hall

Past Worlds, 1988, *Past Worlds. The Times Atlas of Archaeology*, London: Guild Publishing

Peck, J. 1992, 'Labor and agglomeration: control and flexibility in local labor markets', *Annals of the Association of American Geographers* 68: 325–47

Philbrick, A. K. and Brown, R. H. 1973, 'Cosmos and international hierarchies of functions', *Papers of the Peace Research Society (International)* 19: 61–90

Philo, C. 1994, 'History, geography and the "still greater mystery" of historical geography', in D. Gregory, R. L. Martin and G. E. Smith (eds.), *Human Geography. Society, Space and Social Science*, London: Macmillan, pp. 252–81

Philo, P. 1991, 'Foucault's geography', *Environment and Planning, D: Society and Space* 10: 137–61

Pina-Cabral, J. de 1995, 'Paved roads and enchanted mooresses: the perception of the past among the peasant population of the Alto Minho', *Man*, new series 22: 715–35

Polanyi, K. 1957, *The Great Transformation*, with introduction by R. M. MacIver, Boston: Beacon Press. First published in 1944

1968, *Primitive Archaic and Modern Economies*, ed. G. Dalton, Boston: Beacon Press

Pollard, S. 1963, 'Capital accounting in the industrial revolution', *Yorkshire Bulletin of Economic and Social Research* 15: 75–91

1964, 'Fixed capital in the industrial revolution', *Jnl of Economic History* 24: 299–314

1965, *The Genesis of Modern Management. A Study of the Industrial Revolution in Britain*, London: E. Arnold

1985, 'Capital exports, 1870–1914: harmful or beneficial?', *Economic History Review*, 2nd series 38: 489–514

Porter, M. E. 1980, *Competitive Strategy. Techniques for Analyzing Industries and Competitors*, New York: The Free Press

1990, *The Competitive Advantage of Nations*, London: Macmillan

Pounds, N. J. G and Ball, S. S. 1964, 'Core-areas and the development of the European states systems', *Annals of the Association of American Geographers* 54: 24–40

Power, M. J. 1986, 'The social topography of Restoration London', in A. L. Beier and R. Finlay (eds.), *London 1500–1700. The Making of the Metropolis*, London: Longman, pp. 199–223

Pred, A. R. 1975, 'On the spatial structure of organizations and the complexity of metropolitan interdependence', *Papers of the Regional Science Association* 25: 115–42

1977, *City-Systems in the Advanced Economies. Past Growth, Present Processes and Future Development Options*, London: Hutchinson

1984, 'Place as historically-contingent process: structuration and the time-geography of becoming places', *Annals of the Association of American Geographers* 74: 279–97

1985, 'The social becomes the spatial, the spatial becomes the social: enclosures, social change and the becoming of places in Skåne', in Gregory and Urry (eds.), *Social Relations and Spatial Structure*, pp. 337–65

Rapoport, A. 1968, 'The promise and pitfalls of information theory', in Buckley (ed.), *Modern Systems Research*, pp. 137–42

Rappaport, R. A. 1977, 'Maladaptation in social systems', in J. Friedman and M. J. Rowlands (eds.), *The Evolution of Social Systems*, London: Duckworth, pp. 49–71

Readings, W. 1991, *Introducing Lyotard. Art and Politics*, London: Routledge & Kegan Paul

Redfield, R. 1960, *The Little Community and Peasant Society and Culture*, Chicago: University of Chicago Press

Rees, J. 1979, 'Technological change and regional shifts in American manufacturing', *Professional Geographer* 31: 45–54

Reischauer, E. O. 1965, 'The dynastic cycle in China', in Meskill (ed.), *The Pattern of Chinese History*, pp. 31–3

Renfrew, C. 1975, 'Trade as action at a distance: questions of integration and communication', in J. A. Sabloff and C. C. Lamborg-Karlovsky (eds.), *Ancient Civilizations and Trade*, Albuquerque: University of New Mexico Press, pp. 3–59
     1979, 'Systems collapse as social transformation: catastrophe and anastrophe in early state societies', in C. Renfrew and K. L. Cooke (eds.), *Transformations. Mathematical Approaches to Culture Change*, London: Academic Press, pp. 481–507

Renfrew, C. and Poston, T. 1979, 'Discontinuities in the endogenous change of settlement pattern', in C. Renfrew and K. L. Cooke (eds.), *Transformations. Mathematical Approaches to Culture Change*, London: Academic Press, pp. 437–61

Reynolds, S. 1983, 'Medieval *origines gentium* and the community of the realm', *History* 68: 375–90

Rickman, G. 1980, *The Corn Supply of Ancient Rome*, Oxford: Oxford University Press

Ringrose, D. 1989, 'Towns, transport and crown: geography and the decline of Spain', in E. D. Genovese and L. Hochberg (eds.), *Geographic Perspectives in History*, Oxford: Basil Blackwell, pp. 57–80

Robson, B. T. 1981, 'The impact of functional differentiation within systems of industrial cities', in H. Schmal (ed.), *Patterns of European Urbanization since 1500*, London: Croom Helm, pp. 111–30

Roepke, H. G. 1956, *Movements of the British Iron and Steel Industry – 1720 to 1951*, Urbana: University of Illinois Press

Rokkan, S. 1975, 'Dimensions of state formation and nation building: a possible paradigm for research on variations within Europe', in C. Tilly (ed.), *The Formation of National States in Europe*, Princeton: Princeton University Press, pp. 562–600

Rostow, W. W. 1978, *The World Economy. History and Prospect*, Austin: University of Texas

Rowlands, M. J. 1987, 'Centre and periphery: a review of a concept', in M. J. Rowlands, M. Larsen and K. Kristiansen (eds.), *Centre and Periphery in the Ancient World*, Cambridge: Cambridge University Press, pp. 1–11

Rowntree, L. B. and Conkey, M. W. 1980, 'Symbolism and the cultural landscape', *Annals of the Association of American Geographers* 70: 459–74

Rymes, T. K. 1971, *On Concepts of Capital and Technical Change*, Cambridge: Cambridge University Press

Sahlins, M. D. 1968, *Tribesmen*, Englewood Cliffs: Prentice Hall

1977, 'Culture and environment: the study of cultural ecology', in S. Tax and L. G. Freeman (eds.), *Horizons of Anthropology*, 2nd edn, Chicago: Aldine Publishing Co., pp. 215–31

1987, *Islands of History*, London: Tavistock Publications. First published in 1985

Sahlins, M. and Service, E. R. 1960, *Evolution and Culture*, Ann Arbor: University of Michigan Press

Said, E. 1991, *Orientalism. Western Conceptions of the Orient*, Harmondsworth: Penguin Books. First published in 1978

Sauer, C. O. 1952, *Agricultural Origins and Dispersals*, New York: American Geographical Society

Saussure, F. de 1983, *Course in General Linguistics*, ed. and trans. by R. Harris, London: Duckworth

Scammell, G. V. 1981, *The World Encompassed. The First European Maritime Empires c. 800–1650*, London: Methuen

Schama, S. 1987, *The Embarrassment of Riches. An Interpretation of Dutch Culture in the Golden Age*, London: William Collins

Schmid, M. 1982, 'Habermas's theory of social evolution', in H. J. B. Thompson and D. Held (eds.), *Habermas. Critical Debates*, London: Macmillan, pp. 162–80

Schorske, C. E. 1981, *Fin-de-Siècle Vienna. Politics and Culture*, Cambridge: Cambridge University Press

Schrecker, P. 1948, *Work and History. An Essay on the Structure of Civilization*, Princeton: Princeton University Press

Schumpeter, J. A. 1939, *Business Cycles. A Theoretical, Historical and Statistical Analysis of the Capitalist Process*, New York: McGraw Hill

1971, 'The instability of capitalism', in N. Rosenberg (ed.), *The Economics of Technological Change*, Harmondsworth: Penguin Books, pp. 13–42

Scott, A. J. 1988, *New Industrial Spaces. Flexible Production Organization and Regional Development in North America and Western Europe*, London: Pion Ltd

1992, 'The collective order of flexible production agglomerations: lessons for local economic development policy and strategic choice', *Economic Geography* 68: 219–33

Seelan, P. J. M. and Twitchell, D. C. 1974 (eds.), *The Times Atlas of China*, London: Times Books

Shackle, G. L. S. 1961, *Decision, Order and Time in Human Affairs*, Cambridge: Cambridge University Press

1967, *Time in Economics*, Amsterdam: North Holland

Shields, R. 1991, *Places on the Margin. Alternative Geographies of Modernity*, London: Routledge

1992, 'A truant proximity: presence and absence in the space of modernity', *Environment and Planning, D: Society and Space* 10: 181–98

Shils, E. 1975, *Center and Periphery. Essays in Macro-Sociology*, Chicago: University of Chicago Press

1981, 'Center and periphery', in D. Potter (ed.), *Society and the Social Sciences. An Introduction*, London: Routledge & Kegan Paul, pp. 240–54

Silver, M. 1983, 'Karl Polanyi and markets in the ancient Near East: the challenge of the evidence', *Jnl of Economic History* 43: 795–829

Singer, H. W. 1941, 'An index of urban land rents and house rents in England and Wales 1845–1913', *Econometrica* 9: 221–30

Skocpol, T. 1976, 'France, Russia and China: a structural analysis of social revolutions', *Comparative Studies in Society and History* 18: 175–210

1979, *States and Social Revolutions*, Cambridge: Cambridge University Press

Slobodkin, L. B. and Rapoport, A. 1974, 'An optimal strategy of evolution', *Quarterly Review of Biology* 49: 181–200

Smellie, K. B. 1968, *A History of Local Government*, 4th edn, London: Allen & Unwin

Smith, A. D. 1973, *The Concept of Social Change. A Critique of the Functionalist Theory of Social Change*, London: Routledge & Kegan Paul

Smith, J. M. H. 1996, 'Fines Imperii: the marches', in A. McKitterick (ed.), *The New Cambridge History of Europe, vol. II, c. 700–900*, Cambridge: Cambridge University Press, pp. 169–89

Smith, N. 1984, *Uneven Development. Nature, Capital and its Production of Space*, Oxford: Basil Blackwell

1986, 'On the necessity of uneven development', *International Jnl of Urban and Regional Research* 10: 87–104

1989, 'Uneven development and location theory: towards a synthesis', in R. Peet and N. Thrift (eds.), *New Models in Geography*, vol. I, London: Unwin Hyman, pp. 142–63

Smith, W. D. 1985, 'The function of commercial centers in the modernization of European capitalism: Amsterdam as an information exchange in the seventeenth century', *Jnl of European History* 44: 985–1005

Soja, E. W. 1985, 'The spatiality of social life: towards a transformative retheorisation', in Gregory and Urry (eds.), *Social Relations and Spatial Structure*, pp. 90–127

1989, *Postmodern Geographies. The Reassertion of Space in Critical Social Theory*, London: Verso

1996, *Thirdspace. Journeys to Los Angeles and Other Real and Imagined Places*, Oxford: Basil Blackwell

Sorokin, P. A. 1967, 'Reasons for sociocultural change and variously recurrent processes', in W. E. Moore and R. M. Cook (eds.), *Readings on Social Change*, Englewood Cliffs: Prentice Hall, pp. 68–80. First published in 1947

Steensgaard, N. 1974, *The Asian Trade Revolution of the Seventeenth Century. The East India Companies and the Decline of the Caravan Trade*, Chicago: Chicago University Press

1982, 'The Dutch East India Company as an institutional innovation', in M. Aymard (ed.), *Dutch Capitalism and World Capitalism. Capitalisme Hollandais et Capitalisme Mondial*, Cambridge: Cambridge University Press, pp. 235–57

Storper, M. 1985a, 'Technology and spatial production relations: disequilibrium, interindustry relationships and industrial development', in M. Castells (ed.), *High Technology, Space and Society*, Urban Affairs Annual Reviews, vol. 28, Beverly Hills: Sage, pp. 265–83

1985b, 'Oligopoly and the product cycle: essentialism in economic geography', *Economic Geography* 61: 260–82

1992, 'The limits to globalization: technology districts and international trade', *Economic Geography* 68: 60–93

Storper, M. and Scott, A. J. 1989a, *The Capitalist Imperative. Territory, Technology, and Industrial Growth*, Oxford: Basil Blackwell

1989b, 'The geographical foundations and social regulation of flexible production complexes', in J. Wolch and M. Dear (eds.), *The Power of Geography. How Territory Shapes Social Life*, London: Unwin Hyman, pp. 21–40

Sutcliffe, A. 1983, 'In search of the urban variable: Britain in the late nineteenth century', in D. Fraser and A. Sutcliffe (eds.), *The Pursuit of Urban History*, London: E. Arnold, pp. 234–63

Swyngedouw, E. A. 1989, 'The heart of the place: the resurrection of locality in the age of hyperspace', *Geografiska Annaler* 71b: 31–42

Taagepera, R. 1968, 'Growth curves of empires', *General Systems* 13: 171–5

Tainter, J. A. 1988, *The Collapse of Complex Societies*, Cambridge: Cambridge University Press

Thomas, N. 1991, *Out of Time. History and Evolution in Anthropological Discourse*, Cambridge: Cambridge University Press

Thompson, E. P. 1968, *The Making of the English Working Class*, Harmondsworth: Pelican Books. First published in 1963

Thompson, J. W. 1928, *Feudal Germany*, Chicago: University of Chicago Press

Thrift, N. 1981, 'Owner's time and own time: the making of a capitalist time consciousness, 1300–1880', in A. Pred (ed.), *Space and Time in Geography*, series B, no. 48, Lund: Lund Studies in Geography, pp. 56–84

1996, *Spatial Transformations*, London: Sage

Tilley, C. 1981, 'Conceptual frameworks for the explanation of socio-cultural change', in I. Hodder, G. Isaac and N. Hammond (eds.), *Pattern of the Past. Studies in Honour of David Clarke*, Cambridge: Cambridge University Press, pp. 303–86

1975, 'Reflections on the history of European state making', in C. Tilly (ed.), *The Formation of National States in Western Europe*, Princeton: Princeton University Press, pp. 3–83

1979, 'Did the cake of custom break?', in J. M. Merriman (ed.), *Consciousness and Class Experience in Nineteenth Century Europe*, New York: Holm & Meier, pp. 17–44

1984, *Big Structures, Large Processes, Huge Comparisons*, New York: Russell Sage Foundation

1990, *Coercion, Capital and European States AD 990–1990*, Oxford: Oxford University Press

Tilly, R. H. 1978, 'Capital formation in Germany in the nineteenth century', in P. Mathias and M. M. Postan (eds.), *The Cambridge Economic History of Europe, vol. VII, The Industrial Economies: Capital, Labour and Enterprise, part 1, Britain, France, Germany and Scandinavia*, Cambridge: Cambridge University Press, pp. 382–441

Todd, E. 1985, *The Explanation of Ideology. Family Structures and Social Systems*, Oxford: Basil Blackwell

1987, *The Cause of Progress. Culture, Authority and Change*, Oxford: Basil Blackwell

Tosi, M. 1973, 'Early urban evolution and settlement patterns in the Indo-Iranian borderland', in C. Renfrew (ed.), *The Explanation of Culture Change. Models in Prehistory*, London: Duckworth, pp. 429–46

Touraine, A. 1990, 'The idea of revolution', in M. Featherstone (ed.), *Global Culture. Nationalism, Globalization and Modernity*, London: Sage, pp. 121–41

Tuan, Y-F. 1970, *China*, London: Longman

Turner, F. J. 1921, *The Frontier in American History*, New York: H. Holt & Co.

Turner, V. 1974, *Dramas, Fields and Metaphors. Symbolic Action in Human Society*, Ithaca: Cornell University Press

Unger, R. M. 1987a, *Social Theory: Its Situation and Its Task. A Critical Introduction to Politics, a Work in Constructive Social Theory*, Cambridge: Cambridge University Press

1987b, *False Necessity. Anti-Necessitarian Social Theory in the Service of Radical Democracy, Part 1 of Politics, a Work in Constructive Social Theory*, Cambridge: Cambridge University Press

1987c, *Plasticity into Power. Comparative-Historical Studies on the Institutional Conditions of Economic and Military Success. Variations on Themes of Politics, a Work in Constructive Social Theory*, Cambridge: Cambridge University Press

Urry, J. 1985, 'Social relations, space and time', in Gregory and Urry (eds.), *Social Relations and Spatial Structure*, pp. 20–48

Usher, A. P. 1955, 'Technical change and capital formation', in *Capital Formation and Economic Growth*, National Foundation of Economic Research, Princeton: Princeton University Press, pp. 523–50

Vance, J. E. 1971, 'Land assignment in the pre-capitalist, capitalist and post-capitalist cities', *Economic Geography* 47: 101–20

Varaiya, P. and Wiseman, M. 1978, 'The age of cities and the movement of manufacturing employment, 1947–1972', *Papers of the Regional Science Association* 41: 127–40

Vickers, D. 1994, *Economics and the Antagonism of Time. Time, Uncertainty and Choice in Economic Theory*, Ann Arbor: University of Michigan Press

Wacquant, L. J. D. 1992, 'Towards a solid praxeology: the structure and logic of Bourdieu's sociology', in P. Bourdieu and L. J. D. Wacquant, *An Invitation to Reflexive Sociology*, Cambridge: Polity Press, pp. 1–59

Wagner, P. L. 1972, *Environment and Peoples*, Englewood Cliffs: Prentice Hall

1978, 'The themes of cultural geography rethought', *Yearbook of the Association of Pacific Coast Geographers* 37: 7–14

Wagner, R. 1981, *The Invention of Culture*, Chicago: University of Chicago Press. First published in 1975

1986, *Symbols that Stand for Themselves*, Chicago: University of Chicago Press

Wagstaff, J. M. 1978, 'A possible interpretation of settlement pattern evolution in terms of "catastrophe theory"', *Transactions of the Institute of British Geographers* 3: 165–78

Walbank, F. W. 1969, *The Awful Revolution. The Decline of the Roman Empire in the West*, Liverpool: Liverpool University Press

Walker, R. A. 1985, 'Technological determination and determinism: industrial growth and location', in M. Castells (ed.), *High Technology, Space and Society, Urban Affairs Annual Reviews*, vol. 28, Beverly Hills: Sage, pp. 226–64

Wallerstein, I. 1974, *The Modern World-System, vol. I, Capitalist Agriculture and the Origins of the European World Economy*, New York: Academic Press

1979, *The Capitalist World Economy*, Cambridge: Cambridge University Press

1980, *The Modern World-System, vol. XI, Mercantilism and the Consolidation of the European World Economy, 1600–1750*, New York: Academic Press

1987, 'The future of the world economy', in T. K. Hopkins and I. Wallerstein (eds.),

Processes of the World System, Political Economy of the World System Annuals, vol. III, Beverly Hills: Sage, pp. 167–80

1991, *Geopolitics and Geoculture. Essays on the Changing World System*, Cambridge: Cambridge University Press

1993, 'World system versus world systems', in Frank and Gills (eds.), *The World System*, pp. 292–6

Ward, J. R. 1974, *The Finance of Canal Building in Eighteenth Century England*, Oxford: Oxford University Press

Watson, A. 1992, *The Evolution of International Society*, London: Routledge

Watson, W. J. 1959, 'Relict geography in an urban community', in R. Miller and W. J. Watson (eds.) *Geographical Essays in Memory of Alan G. Ogilvie*, Edinburgh: Nelson, pp. 110–43

Weeks, J. 1981, *Capitalism and Exploitation*, London: E. Arnold

Werlen, B. 1993, *Society, Action and Space*, trans. by G. Walls, London: Routledge

Wheatley, P. 1969, *City as Symbol*, inaugural lecture, 1967, London: University College

White, L. A. 1959, *The Evolution of Culture. The Development of Civilization to the Fall of Rome*, New York: McGraw Hill

1975, *The Concept of Cultural Systems*, New York: Columbia University Press

White, S. K. 1988, *The Recent Work of Jurgen Habermas. Reason, Justice and Modernity*, Cambridge: Cambridge University Press

Whitehand, J. W. R. 1972, 'Building cycles and the spatial pattern of urban growth', *Transactions of the Institute of British Geographers* 56: 39–55

1975, 'Building activity and intensity of development at the urban fringe: the case of a London suburb in the nineteenth century', *Jnl of Historical Geography* 1: 211–24

1987, *The Changing Face of Cities. A Study of Development Cycles and Urban Form*, IBG Special Publication no. 21, Oxford: Basil Blackwell

Whitehead, A. N. 1939, *An Introduction to Mathematics*, New York: Henry Holt

1985, *Symbolism. Its Meaning and Effect*, New York: Fordham University Press. First published in 1927

Whitehouse, D. and Whitehouse, R. 1975, *Archaeological Atlas of the World*, London: Thames and Hudson

Whittlesey, D. 1929, 'Sequent occupance', *Annals of the Association of American Geographers* 19: 162–5

Whyte, W. H. 1960, *The Organization Man*, Harmondsworth: Penguin Books. First published in 1956

Wilden, A. 1972, *System and Structure. Essays in Communication and Exchange*, London: Tavistock Publications

Williams, C. E and Findlay, M. C. 1991, 'A reconsideration of the rationality postulate: right hemisphere thinking in economics', in Abell (ed.), *Rational Choice Theory*, pp. 135–54

Wilson, A. G. 1980, *Catastrophe and Bifurcation Theory*, London, Croom Helm

Wilson, C. B. 1992, 'Dwelling at the centre of the world', in E. Lyle (ed.), *Sacred Architecture in the Traditions of India, China, Judaism and Islam, Cosmos*, vol. VIII, Edinburgh: Edinburgh University Press, pp. 111–32

Wilson, J. A. 1949, 'Egypt', in H. Frankfurt, J. A. Wilson and T. Jacobsen (eds.), *Before Philosophy*, Harmondsworth: Penguin Books, pp. 39–70

Winner, L. 1975, 'Complexity and the limits of human understanding', in T. R. La

Porte (ed.), *Organized Social Complexity. Challenge to Politics and Policy*, Princeton: Princeton University Press, pp. 40–76

Withers, C. 1988, *Gaelic Scotland. The Transformation of a Culture Region*, London: Routledge

Wittfogel, K. A. 1957a, *Oriental Despotism. A Comparative Study of Total Power*, New Haven: Yale University Press

1957b, 'Chinese society: an historical survey', *Jnl of Asian Society* 16: 343–64

Wolf, E. R. 1982, *Europe and the People Without History*, Berkeley: University of California Press

Wrigley, E. A. 1962, 'The supply of raw materials in the industrial revolution', *Economic History Review* 2nd series, 15: 1–16

1988, *Continuity, Chance and Change. The Character of the English Industrial Revolution*, Cambridge: Cambridge University Press

Zeeman, E. C. 1979, 'A geometrical model of ideologies', in C. Renfrew and K. L. Cooke (eds.), *Transformations. Mathematical Approaches to Culture Change*, London: Academic Press, pp. 463–79

Ziman, J. M. 1985, 'Pushing back frontiers – or redrawing maps', in T. Hägerstrand (ed.), *The Identification of Progress in Learning*, Cambridge: Cambridge University Press, pp. 1–12

Zukin, S. 1991, *Landscapes of Power. From Detroit to Disney World*, Berkeley: University of California Press

1992, 'The city as a landscape of power: London and New York as global financial markets', in L. Budd and S. Whimster (eds.), *Global Finance and Urban Living. A Study of Metropolitan Change*, London: Routledge, pp. 195–223

Index

Abu-Lughod, J., 54–5, 74
action theoretic approaches, 27–8
Adams, J. S., 158, 173
Adams, R. M., 58–9
adaptation, 16, 24–5, 27, 30, 32–4, 42–3,
    104–5, 134
administration, 67–9, 133, 135, 137, 195–6
    change, 64, 86–91, 137, 195–6
Africa, 194
    north, 68
    trade, 76, 80
    west, 114
Agade, 58
Agnew, J., 81
agricultural innovation, 65
Agricultural Revolution, 195
Albrecht the Bear, 193
Alföldy, G., 68
Allen, P. M., 24, 33–4, 175, 199
America, 98, 121, 155, 171, 194, 198
    Central 74, 76
    industrial regions, 100
    mid-west, 79–80
    Revolution, 187
    South, 74, 76, 80
    War of Independence, 78–9
Amsterdam, 77, 136
Anatolia, 61
Annales school 39
Antwerp, 73–8, 196
AP:PR ratio, 130, 134
Appadurai, A., 166
Arbil, 60
Argentina, 150
Arles, 141
Armenia, 67
Arrow, K., 35, 130–1, 134, 185
Arthur, W. B., 100
Ashby, W. R., 44, 175

Asian societies, 141, 199
Atlantic economy 76, 79, 136
Augustus, Emperor, 68
    Augustan revolution, 68
Aulin, A., 43–4
Australia, 79, 110, 194
Austria, 118
Axelrod, R., 86, 195
    TIT FOR TAT, 86, 195
axial societies, 25
axis mundi, 110, 118

Babylon, 58
Baghdad, 61
Bairoch, P., 74
Baker, A. R. H., 3, 36, 39
Barkow, J. H., 14
Ball, S. S., 91
Baltic, 77–80, 136
bannum lordships, 88
barbarian invasions, 64, 70
Barrow, 148
Bartlett, R., 192–3
Bateson, G., 126, 169, 175–6, 180
Bavaria, 88
Belgium, 98–9
Berger, P. L., 29
Berry, B. J. L., 3–4, 96, 116, 121, 169
Bhabha, H., 10, 17, 44, 165, 181–2, 189,
    199
    énoncé, enunciation, 10, 165
bicameral brain, 13
Birmingham, 90
Black Country, 145
Blaut, J. M., 79, 140, 195
Bock, R. E., 124
Bolivia, 76
Boulding, K. E., 129, 170, 185, 190–1
Boulton, Matthew, 171

Bourdieu, P., 17–18, 104, 108, 113, 119, 137, 169, 175, 177, 201
  concept of capital, 104, 113, 183, 185
  habitus, 17–18, 113
Bowen, M., 93
brain architecture, 14
Brandenburg, 88, 193
Braudel, F., 39, 75, 96, 199
Brenner, R., 38
Bristol, 79, 149
Britain, 67–8, 72, 74
Brooke, Rupert, 28
Brown, R. H., 196
Buckley, W., 24, 27, 43, 47, 127
budget of flexibility, 177
building cycles, 158
built environment, 12, 14, 139–61, 172–3, 179
bureaucracies, 52, 58, 64–5, 89
Burns, R. I., 82
business partnerships, 77–8, 99, 145
Butlin, R. A., 3
Butzer, K., 24, 32, 42, 51
Byzantium, 69–70

Caesar, A., 68
Campbell, B. M. S., 142
Canada, 155
canals, 144–5
Capetian kings, 88
capital, 97, 99, 197–8
  circulating, 143
  contradictions, 9
  depreciation, 146–8, 155
  export, 191
  fixed, 9, 26, 98–9, 102, 139–59, 177, 185, 188
  global, 8–9, 11, 26
  investment cycles, 100–1
  migration, movement, 9, 26, 99–100, 144–6, 156–7, 191–2, 196
  over-accumulation, 191
capitalism, 8–9, 18–19, 26, 38, 47, 54, 70 *et seq.*, 96–102, 197
Carneiro, R., 35
Carthage, 67
Casseti, E., 101
Castells, M., 171–2
catastrophe theory, 24, 49
Chandler, A. D., 82, 97, 136
Changan, 62, 64
change, societal or soci-cultural, 1–2, 16–17, 29–49
  catastrophe, collapse, 24, 31–2, 35–6, 51
  constraint, 36–7, 163, 175–81, 190, 194
  contingency, 7, 31, 164–5, 200
  forms of: evolutionary, 31, 33–4, 131 187–90; involutionary, 31, 46, 184–5; revolutionary, 31, 46, 185–7

geographical displacement, principle of, 46, *see also* 187–95
immanence, 31, 34
life-cycle models, 37–8
morphology of, 47–9
ontogeny, 31, 187
product of, 45–7
qualitative, 46
survival of the fittest, 33–4
taxonomy, 21–50
see also organizational change
chaos theory, 24, 36
Chapman, S. D., 149
Charlemagne, 88
Chaudhuri, K. N., 198–9
Chaunu, P., 192
chemicals, 98, 149
Chi, C., 66
chiefdoms, 36, 49, 85–6, 133, 195–6
Chile, 76
China, 52, 58, 61–6, 82, 187
  empires, civilizations, dynasties; Chang, 63; Ch'in, 64; Ching, 187, 198; Han, 64; Ming, 66; Tang, 64; Zhou, 64
Chomsky, N., 13, 106–7
city-states, 56–8, 66, 93
  Greek, 66, 93
  Sumerian, 56
civilizations, 35, 54, 56–8, 61, 63
Clanchy, M. T., 115
Clarke, D. L., 174
class conflict, 38
classificatory systems, 107–9
Claval, P., 38
Cleveland, 149
cognitive structures, 11, 14
Cohen, S. B., 196
cold societies, 109, 162–3, 168
colonies, 72, 74
colonization, 72, 74, 193
Columbus, Christopher, 000
communications, 53, 64–6, 78, 81
Conkey, M. W., 177
constrained variety, 177–80
Conzen, M., 82
Cook, Captain James, 111–12
core–periphery systems, spaces, 8, 18–19, 52–6, 60–1, 64, 66, 68, 70, 74, 79, 81, 83, 117, 182–4, 188, 191, 193, 195, 197–8, 200
  restructuring, 82–3
Corrigan, P., 89, 92
Corsica, 67
Cosmides, L., 14
cotton industry, 96–8, 149
Cottrell, P. L., 145–6, 149
craft guilds, 129

Crang, P., 120
Crawford, H., 56, 59
culture, 27, 29, 104–22
   identity, 25–7
   local, 26–7
   representations, 25–7
Curry, L., 36–7, 164

Dacia, 67
Danube, 193
Darby, H. C., 2
Dear, M., 102
Denevan, W. M., 30, 106
diachrony, diachroneity, diachronicity, 9–10,
   111, 166–70, 173, 175, 179
Dieterlen, G., 107
diffusion, 49
diminishing returns, 128–9
disjunctures, 165
distributional coalitions, 35–6
Dodgshon, R. A., 46, 74, 107, 113, 142, 193
Domar, E. D., 156, 158
Donald, M., 114
Doyle, M. W., 52, 68
Driver, F., 3
duality, 6, 28
Duncan, J., 118, 178
Durkheim, E., 113
Dutch East India Company, 78

early state modules, 86, 195
East Indies, 74, 76, 78, 144
   East India Company, 144
   see also Dutch East India Co.
economics of flexibility/inflexibility, 180
Edelstein, M., 145
Egypt, 42, 59–60, 67–8
   Egyptian empire, 59–60
Eisenstadt, S. N., 93, 185
Ekholm, K., 38, 45, 53
Elbe, River, 80
Eliade, M., 108, 110–11, 178
elites, marginal, 187
Elton, G. R., 89
Elvin, M., 54, 64, 66
emic-etic debate, 4
empires, 52–70, 140
   Achaemenid, 57, 61
   Akkadian, 57–8
   Assyrian, 57, 60
   Babylonian, 57–8
   British, 80, 187
   Charlemagne, 192
   Chinese, 52, 58, 61–6, 82
   cycles of growth and decay, 54–5, 58–66
   Dutch, 77–9
   Egyptian, 58–60

Holy Roman, 88, 193
Indus, 58
land-based, 93
maritime, 74–5, 78–9, 93
Middle East, 56–61
Mongolian, 66
Ottoman, 54
Parthian, 57, 61
Portuguese, 76–7
Roman, 52, 82
Sasanian, 61
Spanish, 76,
Sumerian, 58
empires, cycles of growth and decay, 54–5,
   58–66
energy, 35, 106–7, 141, 177
England, 90, 121
   administration, 92
   Tudor revolution, 92
Enlightenment, 26, 40, 93, 116
entropy models, 36–7
environmental crises, 24, 32
equilibrium systems, 24, 36
Eridu, 56
Ethiopia, 74
Euphrates, 56–9
Europe, 192, 194, 198–9
   central, 136
   north-west, 97
European Union, 195
événementielle, l'histoire, 39
evolutionary potential, law of, 185
evolutionary theory, 33–4, 46, 49, 125, 175
   evolution and change, *see* change
exchange systems, 33, 38, 53
   unequal, 38, 53–4, 72–3, 75
expended choice, 181, 185, 188, 200
exploration, voyages of, 74–5

Fabian, J., 178
factory system, 149
Faiyum Depression, 60
Falkus, M. E., 150–1
family structures, 95–6
Fanon, F., 165
Far East, 52, 128
Featherstone, M., 26
Feinstein, C. H., 149–53
feudalism, 63–4, 70, 85–9, 92, 192–4
   lords, 88–9, 92, 192–3
   Norman, 192
Flannery, K. V., 128
flexibility/inflexibility, geography of, 180–1,
   183, 186, 197–8
flexible accumulation, 9, 160–1
Florence, 118, 141
Florida, R., 161

Foley partnership, 145
food supplies, 68, 76, 140
Ford, Henry, 167, 171
Fordism, 9, 82, 97, 157, 160, 172
    post-, 9
Forest of Dean, 145
Foucault, M., 7, 104
Fox, W., 93
France, 72, 88, 93–4, 154–5, 181
    capital formation in, 154
Franconia, 88
Frank, G. A., 54–5, 75
Frankish state, 88
Fraser, D., 90
Friedman, J., 38, 53
frontier and marginal societies, 46, 52, 61,
    63–8, 82, 192, 194
Fry, G. K., 154

Galloway, J. A., 142
Ganges, River, 79
Garnsey, P. D. A., 67–71
Gellner, E., 25, 49, 91–2, 94
Genoa, 74–5
    merchants 74
genre de vie, 91, 104
Germany, 67, 81, 88, 96, 99, 127, 148, 154, 193
Gertler, M. S., 101–2, 132, 156–7, 161
Gibson, M., 58–9
Giddens, A., 5–7, 11–12, 16, 28–9, 39–41, 45,
    113, 139, 165
Gills, B. K., 54–5, 69, 75
Gilpin, R., 52–3, 71, 189, 191, 198
Glasgow, 79, 150
global restructuring, 122
globalization, 25–6, 47, 117, 120, 165, 183
gold trade, 76
Goldthwaite, R. A., 118, 141
Goodwin, M., 27
Goody, J., 114–15
Gould, P., 25
Gould, S. J., 171
grain production, 80, 82
    trade, 82
Grassby, R., 140
Great Britain
    net stock of reproducible assets, 152
    gross and net national wealth, 153
    see also Britain
Great Lakes, 78
Gregory, D., 3–5, 13, 16, 40, 50, 96, 190
Gregson, N., 11, 40
Griaule, M., 107
Griffin, T. J., 147

Habermas, J., 4, 14, 27–8, 40, 94–5, 116, 170
    legitimation crisis, 116–17, 170

Haeckel, S. H., 171, 189
Halifax (Nova Scotia), 179
Hall, J. A., 26, 53, 64
Hall, P., 172
Hall, S., 17, 189
Hallpike, C. R., 106, 130, 137
Hanseatic League, 93
Harley, C. K., 80
Harris, C., 3, 5
Harris, M., 4
Harvey, D., 9, 15, 19, 25, 100, 139, 143, 146,
    148, 157–8, 191, 199
Hawaiian society, 112
Hawke, G. R., 145
Hayden, D., 121
Head, L., 110, 178
Hegel, F., 191, 198
    Philosophy of Right (1821), 191
hegemony, 7, 52, 56, 61, 83, 188, 196
    hegemons, 52, 56, 83, 86
Heitland, H. E., 69
Herlihy, D., 141
Hicks, J., 99, 147–8, 173
Hill, C., 93
Hindness, B., 164
Hirschman, A. O., 161
Hirst, P. Q., 164
historicity, 10
Hocart, H. M., 133
Hodder, I., 108
Hoffman, W. G., 154
Holland, 72, 74, 78, 140–1
hooks, bell, 17, 19, 45, 199
Hopkins, K., 68
Höppe, G., 3
hot societies, 109, 162–3, 168
house building, 150, 153–4, 158
    cycles, 150
    invested capital, 152–4
Hsein-yang, 62, 64
Hugill, P. J., 81
human agency, 4, 12, 25–7, 125
Hungarian Plain, 91
hunter-gatherers, 14, 170–1
hydraulic civilization, 56, 58–9, 63–4, 140

Iberia, 67
Ile de France, 91
immigration, 81
imperialism, imperial societies, 190–1
India, 150
Indus Valley, 110
industry, 81, 94, 100–1
    industrial innovation, 97, 99
    industrial organization, 82, 97, 99, 131–3
    Industrial Revolution, 98, 156, 195
    industrialization, 148–59

inertia, 15–16, 35, 39, 84, 87, 102–201
information, 30, 43, 53, 65, 77–9, 81, 114–15, 121, 128, 132, 135–8, 171–2, 174–9, 181, 189, 196, 198
Ingham, G., 81
Innes, M. A., 53, 114
innovation, 63–4, 78, 96–7, 99, 167–8
cycles, 143
intersubjectivity, linguistically produced, 40
investment, 148–59, 173–4
cycles, 148–9, 154, 157–8
primary circuits, 157–8
secondary circuits, 158
involutionary change, *see* change
Iranian Plateau, 61
iron industry, 96–7, 145–6, 149
irrigation, 56, 58–60, 63
Isard, W., 27, 42–3, 46, 157, 170, 176, 196
Italy, 66–7, 93, 118
Ives, B., 172

Jameson, F., 10, 27, 119, 165
Japan, 127
Jarvenpaa, S. L., 172
Johnson, B. L. C., 145
Johnson, C., 123, 182, 186
Johnson, G., 35, 128
joint-stock companies, 144, 157
Jones, A. H. M., 68–70
Jones, E. L., 37, 141–2, 150, 194

Kabyle (Algeria), 113
Kaifeng, 64
Kandyan kingdom, 118
Kaufman, H., 133–4
Keene, D., 142
Keith-Lucas, K., 89–90
Khorsabad, 60
Kuznets, S., 145, 147, 158

labour systems, 54, 65
coercive, slave, 72, 74, 76, 140, 194
corvée, 63, 140
land rents, 158–9
Langton, J., 3
language, 7, 14, 106, 108–10, 115, 126, 169, 175
Larsen, M. T., 115
Lash, S., 123
Laszlo, E., 49
Latium, 68
Lattimore, O., 43, 63–4, 66
Le Play, Frederic, 95
Leach, E., 105
Lefebvre, H., 6–10, 12, 15, 17–18, 157, 182
Levant, 60–1, 67
Lévi-Strauss, C., 107, 109, 111

Lévy-Leboyer, M., 154
Lewis, A. R., 192
Lewit, T., 69
Ley, D., 4, 27, 120
lifeworld, 27–8, 40–1, 92–3, 95, 116–17
liminality, 17, 44
liminal space, 182–3
literacy, 94
Liverpool, 79
Lloyd, C., 11, 39, 104
Lo-yang, 62–3
London, 80, 142, 144, 149, 151, 158
Docklands, 179
great fire, 142
Metropolitan Water Board, 151
longue durée, 39, 112
lordships, 192
Lowe, A., 156–7
Lowenthal, D., 118–221
Lubenow, W. C., 89
Luckmann, T., 29
Lyle, E., 105, 178
Lyotard, J-F., 26

MacDonagh, O., 90
Macedonia, 67
MacKay, A., 82, 194
Madras, 78
Maisels, C. K., 56, 58
maladaptation, 30, 32, 135
Mann, M., 5, 16, 45–7, 53, 61, 189, 198
marginal areas, *see* frontiers
markets, 38, 64, 71, 82, 142
free, self-regulating, price-fixing, 71, 142–3, 173
Markusen, A., 161
marxism, 41, 45
Massey, D., 1, 101–2, 132, 160
Mauritania, 67
Mauss, M., 113
Mediterranean, 142
trade, 74–6
Meinig, D., 2, 74, 76, 81, 119, 187, 194, 198
Merrifield, A., 8
Mesopotamia, 56–61, 67, 110
Mexico, 76
Middle East, 45, 198
Middlesbrough, 148
migration, 94
Mikesell, M., 46
military, armies, 52–4, 63, 67–8, 72, 74–5, 86
Mineral and Battery Works, 144
Missouri, 78
Modernity Project, 40–1, 91, 115
modernization, 91–6, 117
Mokyr, J., 49
moment in being, 168

Mongols, 62, 66
Moors, 76, 194
morphogenesis, 43
morphostasis, 176
Murphy, M., 142
Myer, P. J., 171
Myrdal, G., 100

Nash, D., 67–8
Near East, 52
Nebes, R. D., 13
negentropy, 24, 175
Nelson, R., 33, 35, 41, 49, 131, 190
New England, 100
New World, 194
New York, 81
New Zealand, 79
Newson, L., 46
Nile Valley, 59–60
Nîmes, 141
Nineveh, 57, 60–1
Nisbet, R., 30, 128
Nolan, R. L., 171, 189
non-equilibrium systems, 24
North, D. C., 136, 143, 197
North China Plain, 63

Oates, J., 59
O'Brien, P., 74
Oder, River, 80
Offer, A., 158
Olson, M., 35, 129–30, 179, 185
Ong, W. J., 21, 115, 121
organizational change,
    centralization, 134, 138
    differentiation, 170–1, 178
    entrenchment, inertia, 19, 132–5
    hierarchization, 134, 138,
    innovation, 136–7, 178, 197
    linearization, 134, 138
    plasticity, 19
    specialization, 134, 138, 170–1
organizations, 33–4, 41, 123–38
    see also, socio-political organization

Pacific Rim, 198
Painter, J., 27
Palatinates
    Chester, 89
    Durham, 89
    Lancaster, 89
Paris, 182–3
Parris, H., 89
Parsons, T., 27, 42, 124–5
past, attitudes towards, 118–21
past, rejection of, 8, 10
past–present disjuncture, 10

pastoral societies, 198
path dependency, 161
peasantries, 64, 80
Peking, 62–3, 66
perception, 4, 13, 39
peripheral areas, societies, 16–17, 44–5, 52,
    183, 188–9, 190–1, 199–201
    see also frontiers, marginal areas
Peru, 76
Philbrick, A. K., 196
Philo, C., 3
physical environment, 24, 30, 32, 56, 58–60,
    91, 164
Pina-Cabral, J. de, 167
Poland, 148
Polanyi, K., 71, 97, 116
Pollard, S., 146, 151–3
Pomerania, East, 88
Poor Law Amendment Act, 90
Popper, K., 123–4
population, 74, 90, 109
    collapse, 69
Porter, M.E., 33, 190
Porter, S., 142
Portugal, 74–6
    trade, 74–6
post-modernity, 27, 120, 183
Pounds, N. J. G., 91
Power, M. J., 142
power, power relations, 7–9, 58, 61, 114
Pred, A., 6–7, 100, 164–5
presentational immediacy, 166–7
Preston, P., 172
printing, 94, 177
product cycles, 102
Prussia, 154, 193
Public Health Act, 91
public utilities, 98, 151, 153, 156
    electricity, 98
    gas, 98
    water, 98
Pyrenees, 182–3

railways, 80–2, 136, 144, 150, 153, 156, 179
Rapoport, A., 137, 175
Rappaport, R. A., 33
Readings, W., 26
reconquista, 194
record keeping, 65
recursiveness, 5, 15, 28, 40
Redfield, R., 92
reductionism, 3, 165
Reed, M. C., 145
Rees, J., 100, 102
reflexiveness, 4, 19, 39–40
Reform Act, 90
regulation, 44

Renfrew, C., 36, 46, 49, 85–6, 195–6
resources, 6, 11, 30, 53, 71, 79, 81, 188, 194,
    200
    allocative, 6, 11, 53
    authoritative, 6, 11, 53
    free-floating, 188, 194, 200
    unused freedom, 127, 177, 180, 185, 187,
        191–2
revolutionary change, see change
Reynolds, S., 85, 92
Rhenish League, 93
Rhine, River, 77
Ricardo, D., 158, 173
Rickman, G., 68
Ringrose, D., 76
road construction, 144
Robson, B. T., 101
Roepke, H. G., 101
Rokkan, S., 70, 93
Roman provinces, 87
    towns, 141
Rome, 67–9
Rostow, W. W., 38
Rowlands, M. J., 38, 83
Rowntree, L. B., 177
Russia, 14–15, 187
Rymes, T. K., 147

Sahlins, M., 29, 46, 85, 111–13, 168, 177, 185,
    189, 201
Said, E., 177
Saint Domingue, 78
Saller, R., 67, 69
Sardinia, 67–8
Sauer, C., 2
Saussure, F. de, 109–10, 112, 169
Saxony, 88–9
Sayer, D., 89, 92
Schama, S., 140–1
Schorske, C. E., 118
Schrecker, P., 126, 190, 200
Schumpeter, J. S., 96–7, 143
Scott, A. J., 97, 101, 131–3, 160
Service, E. R., 46, 185
Seville, 76
Shackle, G. L. S., 168
Sheffield, 90
Shields, R., 17, 165–6, 182
Shils, E., 19, 182–3
Sicily, 68
Silver, M., 71
Singer, H. W., 158–9
Skocpol, T., 187
Slavs, 69–70
Slobodkin, L. B., 137
Smellie, K. B., 90
Smith, Adam, 157

Smith, A. D., 42, 125
Smith, J. M. H., 192
Smith, N., 19, 197
Smith, W. D., 78, 136
social theory, 5–10, 15, 16–19
societal or socio-cultural change, see change
societies, 22–9
    complex, 30, 35
    complex adaptive systems, 23, 43
    homeostatic and cybernetic interpretations,
        22–3, 30, 43
    integration, 27, 33, 46, 137
    scalar stress, 45, 47, 69, 128
    steerage, 22, 24, 40, 43–4, 133–4
    structure, 5–6, 10–16, 22–5, 27–9, 106–14
    system-theoretic approaches, 23, 27–8
socio-political organization, 35–6, 49, 85–91,
    128–9, 133–4, 195–6
sociobiology, 12–14, 33–4, 106–7
Soja, E., 2–3, 9, 12, 15, 18–19, 182–3
Sorokin, P. A., 42
South Sea Islanders, 111
space, disordering of, 183
Spain, 72, 76–7, 194
    trade, 74–7
spatial fix, 15, 146, 191–3, 195, 199
spatiality, 2–3, 8, 50, 200
states, European, 196
    proto, 36, 49
steam power, 149
Steensgaard, N., 78
stock exchanges
    Liverpool, 145
    London, 80–1, 144–6
    Manchester, 145
    New York, 81
Storper, M., 97, 132, 160–1
Strabo, 67
structural constraint, 13, 15
structural functionalism, 22–4, 124–5
structuration theory, 6–7, 11, 13, 15, 23, 28–9,
    40, 111–12, 114, 163–4
Sutcliffe, A., 151
Swabia, 88
Sweden, 155
Swyngedouw, E. A., 97, 116, 160
symbolism, 14
synchrony, synchroneity, synchronicity, 9–10,
    111, 166–9, 176, 179
Syria, 67

Tainter, J., 35, 129, 185
Tartars, 66
taxes, 52, 64, 68–9
technology, 33, 37
Tellicherry, 78
temple economy, 56, 58

third space, 17–18, 182–3
Thomas, N., 112
Thompson, E. P., 4, 39
Thompson, J. W., 193
Thracia, 67
Thrift, N., 28, 93, 117
Tigris, River, 56–7, 60
Tilly, C., 181, 196, 199
Tilly, R. H., 154
time–space, 3, 49–50
  compression, 9, 117
  distanciation, 6, 40
  edges, 40, 45
  instantiations, 6
  revaluations, 117
Todd, E., 94–6, 116, 169
Tooby, J., 14
Touraine, A., 185
towns, fire damage, 142
trade, 70–82, 93
  American, 79
  British, 78–81
  Dutch, 72, 79
  French, 78–9
  innovation, 77–8
  long distance, 70–1
  monopolies, 78–9
  West Indian, 78–9
transaction costs, 136
transport, 81, 145, 149, 156
tribes, 195–6
tribute, 56, 67–8
Tudor revolution in government, 90
Turkey, 67
Turner, F. J., 46, 81
Turner, M., 142
Turner, V., 17

Umbria, 68
uneven development, 19, 197
Unger, R. M., 17, 19, 41, 126, 127, 135, 137
  denaturalization, 127
  disentrenchment, 4, 127
  false necessity, 41
  negative capability, 41
  pull of plasticity, 41, 127
  pull of sequential effects, 41, 126, 137, 185
unused freedom, *see* resources
Ur, 56–8
urbanization, 79, 94, 100, 110
Urry, J., 123, 165
Uruk, 56–7
Usher, A. P., 146

Vance, J. E., 139
Varaiya, P., 100–2
Venezuela, 76

Venice, 75, 141, 196
Vickers, D., 172, 201
Vienna, 118–19
vikings, 192
Volga, River, 80
von Thünen, 191

Wacquant, L. J. D., 18, 113
Wagner, P. L., 174, 178
Wagner, R., 112–13
Walbank, F. W., 69
Wales, 89–90
  Act of Union, 89
Walker, R. A., 132, 157, 164
Wallerstein, I., 18, 26, 32, 38, 54–5, 70–2, 79, 197
Ward, J. R., 144, 150
water power, 149, 156
Watson, A., 52–3, 63, 65, 196, 198
Watson, W. J., 179
Weeks, J., 143, 146, 148
Wei-Ho Basin, 63–4
Werlen, B., 25
West Indies, 74–5, 79
Wheatley, P., 30, 110
White, L. A., 41, 162, 174
Whitehand, J. W. R., 150, 158
Whitehead, A. N., 126, 166, 168
Whittlesey, D., 46
Whyte, L., 189
Whyte, W. H., 189
  Organization Man, 189
Wilden, A., 162, 168, 170, 172, 176, 188
Wilson, A., 36
Wilson, C. B., 108–9
Wilson, J. A., 59
Winter, S., 33, 35, 41, 49, 131, 190
Wiseman, M., 100–2
Withers, C., 94
Wittfogel, K., 58
Wolch, J., 102
Wolf, E., 194
world systems theory, 18–19, 32, 38, 52–5, 70–82, 117, 122, 197
Wren, Sir Christopher, 142
Wrigley, E. A., 150, 156
writing, 56, 64–5, 69, 114–17, 128, 177

Xerox Palo Alto Research, 189

Yangtze River, 64
Yellow River, 63

Zagros Mountains, 61
Zeeman, E. C., 36
Ziman, J. M., 178
Zukin, S., 27, 116

Cambridge Studies in Historical Geography

1 Period and place: research methods in historical geography. *Edited by* A. R. H. BAKER *and* M. BILLINGE

2 The historical geography of Scotland since 1707: geographical aspects of modernisation. DAVID TURNOCK

3 Historical understanding in geography: an idealist approach. LEONARD GUELKE

4 English industrial cities of the nineteenth century: a social geography. R. J. DENNIS*

5 Explorations in historical geography: interpretative essays. *Edited by* A. R. H. BAKER *and* DEREK GREGORY

6 The tithe surveys of England and Wales. R. J. P. KAIN *and* H. C. PRINCE

7 Human territoriality: its theory and history. ROBERT DAVID SACK

8 The West Indies: patterns of development, culture and environmental change since 1492. DAVID WATTS*

9 The iconography of landscape: essays in the symbolic representation, design and use of past environments. *Edited by* DENIS COSGROVE *and* STEPHEN DANIELS*

10 Urban historical geography: recent progress in Britain and Germany. *Edited by* DIETRICH DENECKE *and* GARETH SHAW

11 An historical geography of modern Australia: the restive fringe. J. M. POWELL*

12 The sugar-cane industry: an historical geography from its origins to 1914. J. H. GALLOWAY

13 Poverty, ethnicity and the American city, 1840–1925: changing conceptions of the slum and ghetto. DAVID WARD*

14 Peasants, politicians and producers: the organisation of agriculture in France since 1918. M. C. CLEARY

15 The underdraining of farmland in England during the nineteenth century. A. D. M. PHILLIPS

16 Migration in Colonial Spanish America. *Edited by* DAVID ROBINSON

17 Urbanising Britain: essays on class and community in the nineteenth century. *Edited by* GERRY KEARNS *and* CHARLES W. J. WITHERS

18 Ideology and landscape in historical perspective: essays on the meanings of some places in the past. *Edited by* ALAN R. H. BAKER *and* GIDEON BIGER

19 Power pauperism: the workhouse system, 1834–1884. FELIX DRIVER

20 Trade and urban development in Poland: an economic geography of Cracow from its origins to 1795. F. W. CARTER

21 An historical geography of France. XAVIER DE PLANHOL

22 Peasantry to capitalism: Western Östergötland in the nineteenth century. GÖRAN HOPPE and JOHN LANGTON

23 Agricultural revolution in England: the transformation of the agrarian economy 1500–1850. MARK OVERTON*

24 Marc Bloch, sociology and geography: encountering changing disciplines. SUSAN W. FRIEDMAN

25 Land and society in Edwardian Britain. BRIAN SHORT

26 Deciphering global epidemics: analytical approaches to the disease records of world cities, 1888–1912. ANDREW CLIFF, PETER HAGGETT *and* MATTHEW SMALLMAN-RAYNOR*

27 Society in time and space: a geographical perspective on change. ROBERT A. DODGSHON*

Titles marked with an asterisk * are available in paperback.